P9-DNJ-109

Generation
Extra
Large

Generation Extra Large

Rescuing Our Children From the Epidemic of Obesity

LISA TARTAMELLA

ELAINE HERSCHER

CHRIS WOOLSTON

BASIC BOOKS
A Member of the Perseus Books Group
New York

Published by Basic Books
A Member of the Perseus Books Group

Basic Books are available at special discounts for bulk purchases in the United States by corporations, institutions, and other organizations. For more information, please contact the Special Markets Department at the Perseus Books Group, 11 Cambridge Center, Cambridge, MA 02142, or call (617) 252-5298 or (800) 255-1514, or e-mail special.markets@perseusbooks.com.

Designed by Jeff Williams

Library of Congress Cataloging-in-Publication Data

Tartamella, Lisa.
Generation extra large : rescuing our children from the epidemic of obesity / Lisa Tartamella, Elaine Herscher, Chris Woolston.
p. cm.
Includes bibliographical references and index.
ISBN 0-465-08390-0 (hardcover : alk. paper)
1. Obesity in children—Popular works. I. Herscher, Elaine. II. Woolston, Chris. III. Title.

RJ399.C6T37 2004
618.92'398—dc22

2004022121

04 05 06 07/ 10 9 8 7 6 5 4 3 2 1

For our children:

Rachel, William, Angus, and Max.

They always bring out our best.

Contents

Preface

Sometimes important changes have a way of sneaking up on you. One afternoon you're at a busy airport, and you realize that a third of the people waiting for a plane will have trouble fitting into a regular seat. Or perhaps you're taking a stroll through the park, and you see a heavy boy huffing and puffing on his bicycle until he gives up in frustration and walks his bike the rest of the way. Or you're shopping at the mall, and the teenagers seem so much bigger than fourteen-year-olds used to be.

After a while your insights are confirmed when the Surgeon General of the United States reports that both adults and children have become overweight at alarmingly high rates. Worse, they're facing a host of obesity-related illnesses that threaten to shorten their lives.

In 2003, when we first began thinking about this book, the three of us—Lisa Tartamella, Elaine Herscher, and Chris Woolston—were dealing with the nation's obesity epidemic in our own way. In her early years of practicing as a nutritionist at Yale–New Haven Hospital, Tartamella treated very few children who were overweight. Instead, she saw mainly babies and toddlers who were "failing to thrive" and needed special help getting nourishment for their frail bodies. In just a few years, her practice changed dramatically. Now she sees a steady stream of kids, and hardly any are failing to thrive. Her young patients are suffering not from too little food, but from too much.

Elaine Herscher and Chris Woolston—both veteran journalists—had been covering health and medical issues for years when the obesity epidemic began to attract widespread attention. As a former reporter for the *San Francisco Chronicle*, Herscher had written about health policy as well as the AIDS epidemic when it was at its peak. A former staff writer at *Hippocrates*, a Time, Inc., magazine for physicians, Woolston had reported extensively on chronic disease and patient education. Our paths intersected at the editorial department of Consumer Health Interactive, a health website publisher, where Tartamella served as a medical reviewer. While she was treating more and more heavy children and teens, Woolston and Herscher were increasingly focused on stories about kids' sedentary lifestyles, the shortcomings of the American diet, and the fast-food takeover of our country's schools.

The three of us felt compelled to team up and write about this national crisis, but we wanted to approach it in a different way from other books that dealt with the issues facing overweight children. Clearly, the course of the epidemic has been well documented, but something has been missing from the national discourse—a glimmer of optimism. Our book explores the environmental and societal sickness behind childhood obesity, but it also offers practical and urgently needed solutions.

For this book, Woolston and Herscher crisscrossed the country, talking to people in such disparate places as the tiny town of Butler, Alabama; the suburbs of Clinton, Connecticut; the gritty streets of Oakland, California; the urban sprawl of San Antonio, Texas; and the surreal poverty of Las Vegas, Nevada. We visited clinics and hospitals that work with obese children, covered government research panels, talked with worried teachers, and sampled the food at numerous school cafeterias. We attended vigorous debates on junk-food marketing to kids and saw lively demonstrations of a new brand of physical education classes.

Perhaps most important, we talked to the children and teens who are the victims of this epidemic. In the heated debate over childhood obesity, their voices have been all but absent. Many had multiple health problems related to their weight—depression, diabetes, and high blood pressure, to name a few. Some were grappling with poverty as well. A number of overweight kids we interviewed suffered from low self-esteem, but others just wanted to be physically fit and wear regular-sized clothes. From these children, we got the real story of what it's like to be young and overweight in America. From them we also learned what we as adults can do to help them reclaim their health.

In our journey across the country, we heard heartbreaking stories about 100-pound kindergartners and well-meaning parents who lovingly brought their obese youngsters cheeseburgers and fries in place of a school lunch. But we also saw firsthand the heroic efforts of other parents, social workers, nutritionists, physical education teachers, school lunchroom managers, school nurses, and doctors. Their successes against overwhelming odds—a sedentary lifestyle and ubiquitous fast food—should serve as a blueprint for the rest of the nation.

We hope—and trust—that readers will be shocked, disturbed, and angered about what our society has done to encourage childhood obesity. We also hope our book will serve as a call to action, whether it sparks desperately needed leadership on the federal level or inspires changes in one family or one community at a time. Ultimately, we hope we've contributed to a national debate on one of the most destructive epidemics of our time.

LISA TARTAMELLA
ELAINE HERSCHER
CHRIS WOOLSTON

1

An Epidemic for a New Age

FOR WHAT SEEMED LIKE AN ETERNITY, San Francisco General Hospital was at the heart of an epidemic. From the mid-1980s and into the 1990s, people with AIDS came to "the General" by the hundreds and then by the thousands. When the epidemic was at its peak, armies of newspaper, radio, and television reporters trooped through the General, documenting the disease in excruciating detail. Today, Dr. Cam-Tu Tran runs a small clinic at the hospital. There, she treats victims of a quieter but potentially even deadlier epidemic, without fanfare, without much notice at all.

Tran, a small woman in her forties with an expressive face and black hair pulled into a ponytail, is director of an obesity clinic for children, and on this spring day she's coaching an eager medical student through a hypothetical head-to-toe physical examination of an overweight adolescent. "Okay, let's start with the head, what do we look for?" she inquires. "Sleep apnea, thyroid," he says. She nods in approval. "What about cardio?" she says, smiling encouragingly. "The heart rate might sound slower," suggests the student, smiling back. Yes, Tran says, and there may be high cholesterol and early signs of heart disease.

Then Tran skillfully guides the student through the imaginary gastrointestinal tract and ticks off a number of other ailments, among them type 2 diabetes. Moving on to the pelvis, the med student correctly notes that an obese teenage girl might suffer from polycystic ovarian syndrome, a hormonal imbalance that, among other nightmares, can cause a girl to grow thick facial hair. Then Tran briefs the student on a little-known horror faced by obese teenage males: the huge amount of fat in their abdomens may actually hide their genitals, a condition called "pseudo micro-penis," which can cause teenage boys real trauma.

Moving briskly down the body, the doctor flags possible orthopedic problems. An overweight teenage boy could come in with hip pain. "What is your worry right then? Does it mean infection?" Tran asks. The student isn't sure. Tran explains that teenagers who are both overweight and malnourished may actually experience hip fractures from the excess weight. Then, too, some kids are so heavy that their hips literally slip out of their sockets. "You know the term 'slipping'?" she says pleasantly. Tran goes on to explain that surgery is required because the child's enormous weight has put so much pressure on the bone. "You actually have to put pins in," she says.

Seventeen hundred miles to the east, a Texas school administrator is taking her own close look at some overweight students. In her twenty years as a physical education teacher, Gina Castro has seen pretty much every shape of child, but this generation is bigger than any she has taught before. So she decides to weigh every kid in her San Antonio school district. But even from her vantage point deep in the heart of Texas, where heavy children abound, Castro isn't prepared for the results.

In just one school, she finds eight fourth graders who weigh more than 200 pounds. "My God, we had five-year-olds who weighed 100 pounds," Castro says. Although she is aghast at the actual numbers, they are nothing her eyes hadn't already told her. "Twenty years ago, you didn't have a lot of obese kids, and then gradually it was more and more. Then, after a while, I could look down my blacktop with one hundred kids in a line, and thirty of them were obese. I'm not saying just a little overweight, I mean obese."

It is worth wondering what would have happened if Castro had found eight fourth graders in one school with leukemia. For starters, we'd all have heard about it. Teams of epidemiologists would have come barreling into town to study the outbreak. Medical specialists would have begun treating the children and would have reported daily on their condition. Reporters would be locked in combat to be the first to describe just who these nine-year-olds are, how they live, and how they are coping with their heartbreaking disease.

But no busloads of scientists or reporters are sprinting to any of these kids' front doors. It's just another day in the fourth grade.

One of the most disquieting things about obesity is how quickly and accommodatingly we've settled into it, towing our kids along with us. In as little as twenty years we've eaten our way into the record books. Americans now rank among the fattest people on Earth. Two thirds of us weigh more than we should. One third of U.S. adults are slightly to moderately overweight, and another one third are obese. The country's issue with weight caused some 400,000 adult deaths in 2000, according to the Centers for Disease Control and Prevention (CDC), and obesity is poised to overtake tobacco as the leading cause of preventable deaths, possibly as soon as 2005.

As is the case with adults, the current wave of overweight kids is unprecedented in human history. Sixteen percent of U.S. children—approximately one in six—are seriously overweight. That's more than nine million children. Another nine million or so are heavy enough to put their health potentially at risk. Our kids have become extra large with astonishing speed. The percentage of overweight youngsters ages

6 to 11 has tripled since the mid-1970s; it's doubled for teenagers. (For a definition of *overweight*, see page 21)

But by no means is this problem restricted to the United States. Across the globe—from countries as diverse as China, New Zealand, the United Kingdom, even impoverished Uzbekistan—children are ferociously packing on weight. "We're even seeing obesity in adolescents in India now. It's universal," Neville Rigby, policy director for the International Obesity Task Force, told the Associated Press. "It has become a fully global epidemic—indeed, a pandemic."

Old Before Their Time

It would be comforting to see the fattening of the planet's children as simply a social or aesthetic issue, but many overweight kids can't run a lap or pedal a bicycle around the block without pain. Sixth grade may be an exciting and challenging transition to adolescence, but many really overweight kids will remember it as the year they were diagnosed with type 2 diabetes, a disease that used to be found almost exclusively in adults. While our overweight kids anticipate middle school, their blood pressure is slowly creeping up. As they look forward to dating, their arteries are already showing the first clogging streaks of plaque.

These conditions are overtaking kids at younger and younger ages—all as a direct result of being overweight. Because of kids' weight, pediatric blood pressure rates have started to inch up on a national scale, according to a Tulane University study of federal data published in the *Journal of the American Medical Association* (May 2004). The report prompted federal officials to declare that they would be preparing new guidelines for doctors for detecting "prehypertension" in children much as they now diagnose "prediabetes."

"I'm looking at nine- and ten-year-olds and talking to them as if they were fifty-nine," says Barbara King Hooper, a nurse and diabetes educator at Children's Hospital and Research Center in Oakland, California. "None of us was trained to deal with this."

Cheryl's story: "I feel like an old woman"

Cheryl Afghani was only eight years old when she dreamed she had a disease that could kill her. In her dream, the second-grader was walking in the clouds when suddenly she began to fall toward Earth. Just before she hit the ground, she was caught by an angel with white feathery wings and beautiful eyes ringed in blue and gold. The angel told the youngster that she was getting something that was neither good nor bad. She just had to deal with it, and everything would be fine. A few days later, a second dream followed, in which the angel told Cheryl that she had diabetes. The dreaming eight-year-old thought, "I don't have diabetes. My parents have it." Once again she hurtled toward Earth, and once again, the angel swept her to safety. This time the angel had another warning: "This is how fast your life will go if you don't take care of yourself."

Today, Cheryl is a pretty, dark-haired eighth grader with a warm smile. To honor her Pakistani heritage, she wears a gold stud in her nose. She lives with her parents and younger brother in Hayward, California, fourteen miles south of Oakland, in a cramped apartment in a neighborhood of apartment complexes and modest bungalows. At five feet tall, the fourteen-year-old weighs approximately 200 pounds. She was diagnosed with diabetes when she was ten years old, two years after her remarkably prescient dream.

Cheryl's father is from Pakistan, and her mother was born in the Philippines. Both have diabetes. Their daughter is an award-winning singer with a gift for drawing. Her sketchbooks are filled with line drawings of thin young women with big eyes wearing knockout dresses.

"I want to be skinny. I want to fit in those clothes," Cheryl says. "You can talk to anybody, and they'll tell you. They want to be somebody. They want to have that body. They want to feel what the popular people feel," Cheryl says. "And you know what? I have friends, friends who know me, but they are distant to me because I don't look like they do."

Enrolled in advanced placement classes at school, Cheryl has a 3.86 grade point average, and she studies hard. She's more articulate than most teens, and she can do a pitch-perfect rendition of Christina Aguilera's "I Turn to You." But all the pleasure she takes in her talent recedes when she talks about her size. After her diabetes diagnosis, the pounds just began to add up uncontrollably. Cheryl realizes she may eat more than some kids, but she's seen plenty of her thin peers eating pizza and doughnuts with impunity. Lately, she's been walking a lot more, and she thinks it's helping her stamina. By the scale, she hasn't lost weight, but she's hoping that's because she's building muscle. She also says she's managed to cut way back on sweets. "I *loved* chocolate, but now it's like I eat two pieces of a four-square chocolate bar, and I'm full. I'm done."

➔

She's been on every kind of diet; once she even tried skipping two meals a day. More than anything else in the world, she would like to get rid of her extra weight and the rejection she experiences from other kids. When that feeling of shame comes over her, Cheryl says she just cries. "I complain to my mom over and over," she says, tears welling up in her eyes. "And she's like, 'What can I do?'"

Her mother, Eva Afghani, tries to look out for her daughter's well-being as best she can. She lays out her daughter's first insulin shot every morning. She reminds Cheryl she has to love herself and assures her that she's going to lose the weight. She encourages her to exercise whenever she can. But Cheryl's got an exhausting schedule that leaves little room for health concerns. She often comes home with five hours of homework a night. If she can't stay awake late enough to finish, she gets up at 5:30 A.M.—sometimes as early as 4:00—to complete her assignments. By 7:30 she's at school for her job as an office assistant.

Meanwhile, the gym classes at school are geared toward healthy kids who are at a normal weight. Although Cheryl does get some exercise at school, she's invariably the last one to finish the mile run, and in truth she hates it. When she was younger, she used to run up and down the stairs and around her neighborhood all the time, but now her legs tire easily, her heart beats too fast, and she gets dizzy. She remembers the time before diabetes wistfully, the way a middle-aged person might look back on her twentysomething adventures. "Oh, I miss those days . . . when I didn't even know I had diabetes," she says.

Her routine of diabetes care also takes its toll. First thing in the morning, Cheryl checks her blood sugar with a glucose meter. Then she has breakfast and gives herself the first of two daily insulin shots. Then there are her school day, homework, dinner, more insulin, and more homework. She has to watch everything she eats to keep her sugar under control. At the end of the day she's supposed to check her blood sugar one final time, but lately she's just too tired. Between her general fatigue and the effects of her illness, she can rarely keep her head up past nine o'clock at night. "Nowadays my body can't handle it anymore," she says. "I feel like an old woman." •

Until obesity became epidemic, type 2 diabetes was virtually unheard of in children and teens. One study estimates that the number of young victims of the disease has risen at least tenfold in the last two decades. Doctors are making dire predictions about the health of obese youngsters twenty years down the road. But type 2 diabetes is with us right now. It was originally called "adult-onset" diabetes because it seemed limited to people in middle age or beyond. The new generation of patients consists of people in their teens and twenties; some are as young as nine or ten. Faced with a flood of young patients, national health agencies have scrapped the term "adult-onset" altogether.

Dr. Fred Gunville witnessed this transformation firsthand. When he graduated from medical school more than twenty years ago, the young physician had a clear and pressing mission: a type 1 diabetic since childhood, he wanted to treat kids with diabetes. The slim, soft-spoken endocrinologist set up his practice in the early 1980s in Billings, Montana, a quiet town of 90,000 sitting in a valley between the Yellowstone River and a five-mile-long backdrop of sandstone cliffs known as the Rimrocks. The largest town for hundreds of miles in any direction, Billings is a natural hub for medical care. Kids with diabetes come here from hundreds of miles away to see Gunville—from the Crow and Northern Cheyenne reservations, from remote dirt-road mountain towns to the west, and from even more remote railroad and wheat-field towns of the eastern plains.

When he first opened his practice, every kid he saw had the same disease: type 1 diabetes. This disease occurs when the immune system attacks and destroys the factories in the pancreas that make insulin, an essential hormone that helps transport sugar from the bloodstream to hungry cells throughout the body. If the body can't make insulin, the cells starve for energy while the blood becomes overloaded with sugar. For this reason, people with type 1 diabetes need daily insulin shots to keep their bodies nourished and their blood sugar under control.

For years, Gunville witnessed the same pattern: a youngster would come into the office with symptoms that often included fatigue, unquenchable thirst, and too-frequent urination. The kid would be diagnosed with type 1 diabetes and start a lifetime regimen of blood sugar tests and insulin shots. But approximately fifteen years ago, something changed. Gunville started seeing children with type 1 symptoms, but, mysteriously, tests showed they had type 2. Kids with a fasting reading of more than 100 milligrams of glucose per deciliter were considered to have prediabetes. But kids with readings of 126 or more had full-blown diabetes. These kids could make their own insulin, but their blood sugar was still soaring out of control.

The anomaly took Gunville and endocrinologists across the nation by surprise, but it was soon to become routine. Heavy youngsters kept pouring in. For Gunville and just about every other diabetes specialist in the United States, what was once only a trickle of kids with type 2 has become a torrent. "In the last ten years, it really exploded," he says.

Gunville says that every child he has ever treated for type 2 diabetes is overweight, and not just slightly. Some of his patients weigh 300 pounds or more. The association between obesity and diabetes is no mere coincidence. No matter what a person's age, extra pounds can upset the body's delicate chemistry. As fat cells grow larger, they release an array of compounds such as leptin that block the action of insulin. As a child grows fatter and fatter, insulin may lose more and more of its punch, a phenomenon that doctors call insulin resistance. Sugar will slowly build up in the bloodstream, creating a toxic sludge that can damage cells all over the body. The body compensates by producing more insulin, but even that may not be enough. Eventually, the blood sugar can climb higher than normal—the hallmark of prediabetes. If something isn't done to restore insulin's power, blood sugar will continue to rise until a person reaches full-blown diabetes.

In the short term, this condition can cause blurred vision and can slow the healing of wounds. In an unfortunate twist, extra sugar also harms the cells in the pancreas that produce insulin. The insulin is not only much weaker than before, but it also becomes much more scarce. Sugar levels continue to climb, and the sludge begins to eat away at the kidneys, the retinas, and the nerves and blood vessels in the extremities. William Dietz, MD, PhD, a leading obesity researcher with the CDC, predicts that many children diagnosed with the disease today will start to suffer the worst of its complications in their twenties or thirties. That means they can anticipate kidney failure, heart disease, blindness, and amputations at the point when they're launching a career or starting a family.

Gunville knows what it's like to prick the same finger over and over again to test his blood sugar levels, and what it's like to gnaw on an apple when the whole world seems to be enjoying candy and ice cream. Still, when kids with type 2 diabetes come into his office, he says he often sees sadness and despair far beyond his personal experience. Gunville's roster of kids with type 2 reads like a roll call of the downtrodden. This one lives in a group home, that one is severely depressed, yet another basically fends for himself every day. In some cases, Gunville says, kids suffer mental anguish because they are overweight and battling a chronic disease. For other kids, obesity and diabetes seem to have stemmed from their emotional problems. Either way, these kids face obstacles that would make most adults give up.

A thousand miles away in a highly urban setting beset by high levels of poverty and crime, Dr. Suruchi Bhatia has had exactly the same experience. Many of her patients are so overwhelmed by other things in their lives that obesity and diabetes are just two more miseries to wake up to. "The sense I get from them is that they never really made it out of the starting gate and just feel like they wish they could start over," says Bhatia, chief of the Division of Diabetes and Endocrinology at Children's Hospital and Research Center in Oakland, California. "I see kids, and I talk to them about the complications, and they say, 'I don't care. It doesn't matter anymore.' They feel like there's nothing they can do to get back to a healthy state."

Children with the least access to medical care are also the most likely to get type 2 diabetes, according to Bhatia. In fact, federal data show that the disease disproportionately affects the poor and the nonwhite. Particularly vulnerable are Latinos, African Americans, some Asians and Pacific Islanders, and Native Americans. "I think I couldn't count for you one Caucasian type 2 diabetic [in my practice]," says Bhatia. In that city's school district, administrators say they feel as though they're watching a tsunami headed their way. The district estimates that up to thirty-eight percent—as many as 21,000—of its 54,000 students are at risk for type 2 diabetes based on their weight and ethnic origin.

The CDC predicts that fully one third of those born after 2000 will have type 2 diabetes at some point in their lifetime. (For Latino or African American kids born at that time, the odds may be closer to one out of two.) That's one third of a generation living with a potentially deadly disease brought about, for the most part, by too little exercise and too much food.

Diabetes is by no means the only threat. Other experts are noting grimly that today's children may be the first in history to have a shorter life expectancy than their parents. Studies show that severely overweight children are likely to become severely overweight adults. A twenty-year-old who is very overweight may expect to live thirteen fewer years than someone the same age who has a normal weight, according to a study from the University of Alabama at Birmingham. "It's a major public health problem," says the study's author, David P. Allison, PhD, "and years of life lost is just one consequence." As a nation, we're spending approximately $117 billion a year on obesity-related health issues. From 1979 to 1999, obesity tripled in some age groups, and the cost of treating obesity-related conditions in kids tripled as well.

Who's to Blame?

In conversation about the problem of overweight kids, someone invariably will volunteer a tale about the kids in his teenager's class who are too big to run a half mile or how her cousin's bridesmaid's nine-year-old daughter is so obese she couldn't fit into a dress for the wedding. Inevitably, that person will say, "What are the parents thinking? How could they let them get like that?"

Many parents of overweight kids are asking themselves the same thing. In a recent A. C. Nielsen poll, two thirds of parents of overweight children blamed themselves. Another ten percent blamed the child. The results are hardly surprising. Many Americans—including many overweight Americans—believe obesity is a direct result of poor choices and weak willpower. In a Time/ABC News poll reported in the summer of 2004, eighty-seven percent said that individuals were to blame for the nation's obesity problem. Even Tommy Thompson, the pedometer-toting secretary of the Department of Health and Human Services under George W. Bush, has said that the key to preventing and treating obesity is "personal responsibility." If responsibility is the answer, irresponsibility must be the problem.

But that's not what legions of doctors, nutritionists, parents, teachers, and public health officials think. Yale University psychologist Kelly Brownell, co-author of *Food Fight,* reasons that our penchant for fast, high-calorie food and supersized portions, coupled with a precipitous decline in exercise, has led to a "toxic environment" that essentially overwhelms people—especially if those people are children. Brownell says that a full-fledged public sector response is slow in coming in part because of the federal government's "personal responsibility" mantra.

"Obesity has been considered a consequence of weak discipline, psychological dysfunction, and other personal failings," he writes. "It is widely believed that obese people are responsible for their condition and that they—not physicians, insurance companies, or the nation—should be responsible for its remedy." Brownell emphatically disagrees. Children in particular, he says, "need protection from a food and activity environment that is out of control."

One problem is that the U.S. food industry produces 3,800 calories a day for every man, woman, and child in the United States, says Marion Nestle, New York University nutritionist and author of *Food Politics.* Women need approximately 2,200 calories, men approximately 2,500, and young kids less than 2,000. Food manufacturers can't sell all those extra calories without increasing portions and seducing people to eat more, she says. "There's something about human psychology—if a lot of food is put in front of us, we eat it," Nestle, an outspoken critic of the food industry, said in an interview with *Fortune Magazine*. "People who believe our eating habits are a question of individual free will are much, much stronger than I am."

Radley Balko, a policy analyst with the libertarian Cato Institute and a speaker at a major obesity summit in June 2004, says he couldn't disagree more. He has support from many quarters, including the halls of government, when he argues that obesity never should have been a public health issue at all. "Instead of manipulating or intervening in the array of food options available to American consumers, our government ought to be working to foster a sense of responsibility in and ownership of our own health and well-being," Balko wrote in an article for the Internet. "We're becoming less responsible for our own health and more responsible for everyone else's. Your heart attack drives up the cost of my premiums and office visits. And if the government is paying for my anti-cholesterol medication, what incentive is there for me to put down the cheeseburger?"

That reasoning is argued vociferously by the food and beverage industries and reflected again and again in people's attitudes about weight, notes Joan Miller, who runs an antiobesity campaign in San Antonio. "Especially in places like Texas, there's more of this attitude of self-sufficiency," Miller says. "It's like, well, if you're fat, take care of it yourself. It's a moral issue. It's 'you're lazy, and you're not caring for yourself, so why should I care about you?'"

Clearly, people are responsible for what they eat, but they can't be held accountable for keeping a level playing field between themselves and the food industry. Doing so is especially difficult when the fast food industry alone spends $3 billion a year in advertising to children, nearly 1,000 times the federal government's $3.6 million yearly budget promoting consumption of fruits and vegetables.

The playing field isn't so level outside the United States anymore either. We may be the leader in consumption of high-calorie food, and we may have invented the sedentary automobile-driven culture, but we have plenty of followers. Kids in Beijing, Stockholm, Boston, and Cairo are beginning to have more and more in common. They are surrounded by cheap, tasty, high-calorie food. They spend much of their time in front of TV and video screens. Their parents have little free time to help them exercise and eat right. And they're getting fatter. Heavy kids are products of an environment that is pushing them to overeat and stay as inactive as possible.

Being overweight is nothing more than an imbalance between what we eat and the energy we expend, but beyond that basic equation, nothing about this epidemic is simple. For Americans, in fact, it goes straight to the heart of what's wrong with our culture. Obesity is in large measure the embodiment of the way we live. Our physical beings have become extra large billboards for some of our most pressing social problems.

Soda for Breakfast

Common wisdom says we are too fat because we have too much. Too many food choices in the supermarket. A drive-through on every other block. A soda machine in the hallways of practically every high school. Experts agree that the abundance of food choices, especially the inundation of high-calorie, high-fat food, has had a profound impact on our culture and way of eating. But in another sense, this is an epidemic of poverty and need as well as abundance.

The majority of overweight kids are fat not because they have too much but because they have too little. The highest rates of obesity occur among the poorest children. In 2002, thirty-five million Americans lived in households in which there wasn't enough money for food on a consistent basis, according to the Food Research and Action Center. Those are the very people most likely to be obese. In 1998, twenty-six percent of adults with incomes below poverty level were obese, compared to fifteen percent of adults in the highest income brackets.

As with any public health issue, what affects the health of the parents also affects the health of the kids.

In a study of low-income Latino families in California, more than three fourths of the women were overweight or obese; twenty-two percent of their kids were overweight as well. Yet among sixty percent of the households studied, there was a period during each month when they couldn't afford nutritious food. When researchers studied low-income preschoolers in Head Start programs in California, they found that thirty-two percent were overweight. "Overweight has replaced malnutrition as the most prevalent nutritional problem among the poor," says Patricia Crawford, co-director of the Center for Weight and Health at the University of California at Berkeley. The reasons for the seemingly contradictory hunger/obesity link aren't hard to figure out. Crawford's UC Berkeley study found that in times of scarcity, such foods as bananas, yogurt, and tomatoes were gone, but less expensive Kool-Aid, hot dogs, and sugar-sweetened cereals remained diet staples. "If you're a poor person and you've got $5, and you can take that $5 to McDonald's and buy five hamburgers or one salad, which will you choose?" asks NYU's Nestle.

When Terre Logsdon of the West Oakland, California, YMCA began helping elementary schools run their physical education programs, one of the first things she noticed was the students who were too hungry to exercise. Twenty percent of her students were so overweight they were embarrassed to get out on the field, but the underlying truth was that many of them weren't eating well enough to muster the energy to run and play. Those are the kids who come to school with a bag of Cheetos and a bottle of Gatorade for lunch, says Logsdon. "The students don't do well in PE because they're hungry. It's such a vicious cycle," she says.

Stressed to the Max

Public health experts also see a direct link between gun violence and obesity in the inner cities. To be a poor kid living in a dangerous neighborhood is a bit like being a prisoner. In environments where gun violence is a regular occurrence, many parents feel more secure with their kids sitting safely inside watching TV. In Oakland, diabetes educators often struggle to devise a physical activity plan for their young patients. The best they can do sometimes is advise a mother to walk her preteen daughter to the local liquor store and back just to get a bit of exercise.

No income bracket is immune to the new epidemic. Our culture seems to be perpetually running on fast-forward, and many middle- and even upper-middle-class children and their parents are overcommitted and undernourished. In the majority of two-parent households, both parents are working outside the home just to maintain a house in a safe neighborhood with good schools. Meanwhile, the pace of work becomes ever more demanding, gobbling up family and leisure time. According to the Bureau of Labor Statistics, more than twenty-five million Americans work more than forty-nine hours a week; eleven million of them work more than sixty hours weekly. In her book *White-Collar Sweatshop*, Jill Andresky Fraser makes the case that with the shrinking job market and the unstable economy, more Americans are forced to work harder than ever. Two paychecks, she says, "mean twice as much potential for overwork and exhaustion." Parents may be able to afford organic micro greens and free-range chicken, but who has the time and energy to buy it, prepare it, and clean up the kitchen afterward?

Even some of the nation's leading nutritionists can't always pull a meal together after a long workday. "I have enough resources financially," says Joan Carter, a dietitian who teaches in the Department of Pediatrics at Baylor College of Medicine. "I have a car to get me back and forth to work and a grocery store around the corner, and I still have trouble doing it some days." As family leisure time has evaporated, so has the family meal. And fast food franchises have eagerly sprouted everywhere to fill our insatiable need to eat and run.

This dynamic wouldn't work so handily against kids if they were literally eating and running. But for many kids running is not a vital, natural part of childhood any more—neither is any other form of exercise. The exercise that came to many of their parents organically—walking to school, biking around town, just running free in the neighborhood—is rarely an option for today's kids.

The typical American neighborhood, middle class or not, is considered unsafe for young children's play, either because of the real level of crime and violence or the perception that all children are potential targets for predators. Kids may also be at risk riding their bikes or crossing the street in high-traffic areas. In 2001, the National Safe Kids Campaign found that nearly sixty percent of all kids have to overcome at least one "serious physical hazard" on their way to school. So children's exercise is often scheduled with an adult manager present—usually a stressed-out parent who's juggling too many responsibilities and is having trouble getting exercise herself.

Doctors who treat overweight kids have firsthand experience of how time pressure and other stresses manifest in their patients. Bhatia has time and again observed the toll on her young patients from breakneck schedules, unhealthy food choices, and limited opportunities to be physically active.

"You know, I read about people in France having eight weeks of vacation and essentially everything shutting down in the summer, people spending time with their families and I think wow, what a wonderful place," says Bhatia, whose family is from India. "I think for our level of wealth here in this country, we lead relatively unhappy lives."

That level of wealth isn't apparent in most American schools, either. The connection may not be transparent, but the budget problems of schools have had a direct effect on children's waistlines. Schools are crumbling structurally and are in such poor shape financially that one in three teachers reported in 2002 not having enough textbooks to send work home. Stretched to the limit financially, schools have turned to money-making fast food purveyors—McDonald's, Taco Bell, KFC, and Pizza Hut among them—to feed their students at lunchtime. To ease their budget problems, more than three quarters of all U.S. schools have entered into exclusive, often secret, vending machine contracts with Coca-Cola and Pepsi, putting soda within easy reach of young children and teens throughout the day.

The Department of Health Services advocates parental responsibility but overlooks conditions in schools that completely undermine parents. "When we drop off kids at school, we expect them to be taken care of," says Margo Wootan, nutrition policy director at the Center for Science in the Public Interest in Washington, D.C. "Parents aren't at school. How are they supposed to monitor what their kids are eating?" she says. The school food environment is clearly "much worse" than it was a generation ago, she says. "It's not that kids didn't have treats twenty years ago. It's that now they have treats all the time."

Kids' bodies might be able to tolerate their schools' "international junk food courtyards," in the words of one school food service director, if only they got enough exercise at school to balance it out. But in today's school economy, physical education is considered frivolous. The pressure to improve academics has squeezed out physical education entirely in some places. Only eight percent of elementary schools and six percent of middle schools and high schools provide daily phys ed. Many have even cut back on recess time, too, despite the fact that studies show physical activity helps kids perform better on tests.

New research now shows that an environment of poor nutrition and exercise for children is costing schools million of dollars each year. Because of health problems, overweight kids tend to be absent from school more often, and each time they call in sick, their school districts lose funding. According to a report released in September 2004 by the nonprofit Action for Healthy Kids, a large city such as New York could be losing $28 million each year. When the nonprofit, founded by former Surgeon General David Satcher, analyzed the data on overweight kids and the rate at which they are absent, it found that an average-size school district could be forfeiting between $90,000 and $160,000 annually. The study also found that obesity influences how well students perform on tests. Schools with a high percentage of kids who do not eat well or exercise regularly reported lesser gains in test scores.

Overweight and Depressed

Overweight kids don't worry much about heart disease or cancer or other distant threats. For them, the present is just about all they can bear. They get called names on the playground. They get picked last in gym. They have trouble making friends. They are bullied or they turn into bullies themselves. One recent survey done by researchers at the University of California at San Diego found that overweight kids have a worse quality of life than young cancer patients.

That's been fifteen-year-old Yadira Renteria's experience. All her life, kids have called her names. Even now that Yadira is almost in high school, the teasing still hasn't eased. "They call me fat. They call me a pig. And there are some boys that call me Big Mac. And they call me Big Mama," she says, barely able to get the words out. "If I'm walking on the street they start screaming. They don't care if they hurt your feelings."

Several times Yadira has told her worried mother that she'd like to die. Along with weighing 260 pounds, Yadira has type 2 diabetes. "She was eating salads, walking, but her weight really wasn't going down. And she doesn't understand. She says, 'Why?' Why doesn't her weight go down?" her mother, Patricia Orozco, says in Spanish. "I see that every day when the sun goes down, she goes out to walk. She walks one or two hours, and her weight doesn't go down." Along with exercising, Yadira is following a nutritionist's eating plan, but her mother says it hasn't helped. "My girl has done all of this, but . . . I see that my girl is a little fatter. I don't know what to do.

"I tell her, 'Child, look, you're pretty. (The kids who tease you) are just jealous.' I tell her, 'You're young. I can help you, honey. Get this out of your head, out of your mind that you don't want to live because I love you very much, sweetie.'" Orozco says that her husband also has diabetes but that his is well managed. As Yadira's weight and her blood sugar levels climb, her emotional state becomes more fragile. The teenager was about to have a formal graduation from middle school, and even that happy occasion caused her pain. "She got sad because she has to wear a gown, and it was too small. But I'll make her gown myself," her mother says. "And she'll be very pretty."

Even as heavy kids become more commonplace, the stigma attached to obesity has only deepened. Researchers from Yale University showed 458 fifth and sixth graders pictures of six different children, including one child in a wheelchair, one with a facial deformity, one who was overweight, and one who looked "normal." The study was an exact replica of one conducted in 1961, but the results were dramatically different. When the kids were asked who was more likable, the overweight kid fared far worse today than forty years ago.

By querying 106 obese children and their parents, pediatric specialists at the University of California at San Diego found that the kids' quality of life was far inferior to average-weight kids. They were five times as likely to be impaired physically and to be suffering far more than other youngsters emotionally, socially, and at school. Approximately sixty-five percent of the kids had an obesity-related health problem, such as diabetes, sleep apnea, or elevated cholesterol; nearly thirteen percent suffered from anxiety, depression, or other psychiatric problems.

Many kids respond to the ridicule with self-loathing. They believe that they must, at all costs, be skinny. And their shame rises with the needle on the scale. In the magazines with the anorexic cover models, the advertising inside holds out the hope of extreme thinness with page after page of either useless or dangerous weight loss concoctions. Some overweight kids go on to develop anorexia and bulimia, eating disorders that have struck tens of thousands of American teenagers.

Indeed, children are becoming increasingly weight conscious at a younger age. In a Southern Illinois University survey of more than 1,100 nine- to thirteen-year-olds, fifty-nine percent of the kids said they have tried to lose weight. Even half the kids who said they were at about the right weight had tried to lose. And more than forty percent of kids who said they were "slightly" or "very" underweight had been on a diet. But on a note of hope, fully sixty-nine percent of the kids said that "eating healthy and exercising" were the best ways to control body weight, finishing way ahead of the seventeen percent who thought that "going on a diet" was the best solution.

If kids are overly self-conscious, it's often their parents who are in denial. In a study of 622 mothers of different income levels and ethnic backgrounds, nearly all obese mothers believed they were overweight. But in a fascinating testament to the blindness of maternal love, only one in five correctly identified their overweight children as having a problem.

Melanie Ritsema, director of a supplemental food and nutrition counseling program in San Antonio, comes across parental denial every day, but she recognizes why parents dismiss the problem. Developing a major health problem from obesity can take many years. "It's such a long progressive thing, and we're into such instant feedback, that I just don't think we can see it," says Ritsema. "We don't draw a connection with this long process of gradually poisoning ourselves."

Nobody believes that all kids need to fit into a cookie-cutter model of thinness. In fact, a growing movement among health professionals instead touts the philosophy of "health at every size." A new breed of exercise physiologists, academics, physicians, and nutritionists supports the idea that if you aren't morbidly obese, weight loss isn't nearly as important as being physically fit and getting the right amount of healthy foods every day. Doctors suggest that even kids who are extremely overweight should just try to hold their weight steady as they grow taller.

Regardless of weight, all kids need to hear the same message: they need to eat well and stay active. When health experts envision a better future, they don't see a bunch of little stick figures. They see kids running in parks and playgrounds, riding bikes around the neighborhood, and playing tag in their backyard, all the while fueled with healthy foods. Kids would still come in different sizes, but they wouldn't be too heavy to walk down the block, and they wouldn't be old before their time.

People like Melanie Ritsema, Joan Miller, Cam-Tu Tran, Fred Gunville, and Suruchi Bhatia are working in communities across the nation, on a deeply personal level, to help people make those connections and to begin to turn this epidemic around. U.S. Surgeon General Richard Carmona, who calls the obesity epidemic "the terror within," summed up these efforts in 2002 when he addressed a gathering of experts seeking to bring healthy changes to schools: "We need to lead a cultural transformation." Neither Carmona nor anyone in the grassroots has any illusions about how hard that's going to be. But they do know that what they do now may shape our health as a nation for generations to come.

Body Mass Index

Health practitioners commonly gauge overweight and obesity with a measure known as the body mass index or BMI, a formula that takes both weight and height into account. (To calculate BMI, take weight in pounds, divide by height in inches, divide again by height in inches, and multiply the result by 703, or, even better, search for "BMI calculator" on the Web.) Adults with a BMI between twenty-five and thirty are considered overweight, and a BMI over thirty makes them officially obese. Because the shapes of kids' bodies change so dramatically as they grow, those same rules don't apply to children. A child with what looks like a low BMI by adult standards (a BMI of 17, for example) may actually be overweight, depending on his age.

Doctors use special growth charts to see if a child's BMI is high, low, or about average for his or her age. By the definition of the CDC, kids are officially overweight if their BMI puts them at or above the ninety-fifth percentile on the growth charts. In other words, their BMI would be higher than ninety-five percent of kids

➔

of the same age and gender in the early 1970s, when the charts were first constructed. The CDC says all kids between the eighty-fifth to ninety-fifth percentiles are "at risk for overweight," but this label is controversial. The International Obesity Task Force believes that kids in the eighty-fifth percentile should also be considered overweight because they face weight-related long-term health risks. To complicate matters a little further, many health practitioners use the term "overweight" for kids in the eighty-fifth to ninety-fifth percentiles and "obese" for kids at or above the ninety-fifth percentile.

Table 1.1 Body Mass Index Chart

Examples of kids at the 85th percentile for BMI

BOYS

Age	*Height (feet and inches)*	*Weight (lbs)*	*BMI*
5	3'7"	44	17
10	4'6"	81	19.5
15	5'8"	155	23.5

GIRLS

Age	*Height (feet and inches)*	*Weight (lbs)*	*BMI*
5	3'7"	44	17
10	4'6"	83	20
15	5'4"	140	24

Examples of kids at the 95th percentile for BMI

BOYS

Age	*Height (feet and inches)*	*Weight (lbs)*	*BMI*
5	3'7"	47	18
10	4'6"	91	22
15	5'8"	178	27

GIRLS

Age	*Height (feet and inches)*	*Weight (lbs)*	*BMI*
5	3'7"	48	18
10	4'6"	95	23
15	5'4"	168	28

SOURCE: CDC Body mass index-for-age percentiles, 2000 (some numbers have been rounded)

2
Fat City

SHORTLY AFTER DAYBREAK on the morning of March 6, 1836, General Antonio López de Santa Anna's 2,400 Mexican troops swarmed the Alamo, killing every soldier inside, for a time returning control of the city of San Antonio to Mexico. Among the dead were Davy Crockett and Jim Bowie, legends who helped shape the memory of the Alamo massacre as a war of men against overwhelming odds. Today the city of San Antonio is yet again at war for its future—only this time, the casualties are dying slowly, not from cannon fire, but from food.

Known alternately as the Alamo City or the Fiesta City, San Antonio often goes by another, less festive name: Fat City. By the federal government's account, San Antonio is the fattest city in America. In a 2001 survey, the Centers for Disease Control and Prevention found that Alamo City had the highest percentage of obese adults of any American metropolis. A widely publicized study from *Men's Fitness* magazine rates San Antonio farther down—fourth in fat adults behind Detroit, Houston, and Dallas. But by anyone's numbers, San Antonio belongs in the pantheon of the extra large.

Spend a little time in San Antonio, and the city's problem becomes evident: an overweight mom and dad and two heavy kids strolling near El Mercado, the Mexican marketplace; pudgy kids piling off the bus for school; children in extra large T-shirts trudging through gym class; a man holding the hand of his grandson, the squat, chubby toddler tugging on Grandpa to go the other way. Indeed, sixty-five percent of the city's adults are overweight; from twenty-five to thirty-one percent are obese. In some neighborhoods, as many as seventy-six percent of residents are overweight.

In San Antonio, partying and recreational eating have been elevated to an art form. Fiesta time in late April, a ten-day block party and food fest, offers the city's biggest excuse for indulgence. Fiesta is so significant that business comes to a standstill and city workers get a day off. Along with the parades and the music, the streets are lined with food stalls selling fajitas, enchiladas, gorditas, funnel cakes, corn dogs, and sausages. Here you can purchase a whole turkey leg to munch on while you watch the Parade of Flowers go by. And even when it isn't Fiesta time, the city's food binge never ceases. "Ask anyone in town 'Where's the buffet?'" says nutrition counselor Anna Guerrero. "They'll know."

But while the rest of America is just waking up to the national obesity epidemic, San Antonio has spent the last six years doing something about it. In an unprecedented move, San Antonio has mobilized every key player in the city—political leaders, public health officials, doctors, school food service directors, phys ed teachers, social workers, and parents—in its Fit City/Fit Schools initiative. In 2003, the Congressional Hispanic Caucus Health Task Force declared San Antonio's antiobesity campaign "a model for the rest of the nation."

San Antonio's health department got its first wake-up call in 1998, with a citywide health assessment of everything from teen pregnancy to mental health, obesity, and related diseases. The health department found that obesity was climbing, type 2 diabetes was soaring, and heart disease was responsible for a third of the city's deaths. By their self-reports, nearly half the county's residents were sedentary. Faced with these numbers, San Antonio's hospital systems, normally engaged in cutthroat competition, threw down their scalpels and decided to work together and jointly fund a communitywide antiobesity campaign. "It was too critical to avoid any longer," says Joan Miller, executive director of what would become the Bexar (pronounced "bear") County Community Health Collaborative.

It's not glamorous work. It involves convincing food companies they can still make money if they stock vending machines with healthier food. It means cajoling residents to join a walking campaign in a city that's hot for most of the year and about as pedestrian-friendly as a NASCAR racetrack. It requires challenging some parents' most sacrosanct food traditions and teaching them a healthier way to feed their kids. It means supporting school nutritionists in replacing sugary and fatty foods in the cafeteria even if they lose money. It means instituting a new breed of physical education classes while being pressured from all sides to concentrate on academics.

One man's dedication is already paying off. From his small nonprofit program in one of the city's poorest neighborhoods, Roberto Treviño is fighting type 2 diabetes in children and accomplishing what Ivy League research departments have not. One of the first doctors to notice the epidemic of diabetes in young children, he has lowered dangerously high blood sugars in low-income nine-year-olds by changing their eating and exercise behaviors. Treviño is doing this without complicated drug regimens—just diet and exercise—almost surely saving these children from a lifetime of diabetes. Much of the rest of the collaborative's grassroots work won't yield tangible results for years, but the hope is to transform San Antonio from Fat City to Fit City, one person at a time.

Rich Kid, Poor Kid

With 1.1 million people, San Antonio is the ninth-largest city in the nation, but it operates like a small town. "Everybody knows everybody," says Miller. Despite its proximity to the Texas hill country, the city is mostly flat, so flat that multistory buildings can be seen for miles as you approach on one of the many interstates that cut through the city's heart. Most of the major highways in Texas meet in San Antonio: I-10, leading to El Paso to the far west and Houston to the east; I-35, connecting San Antonio to Austin, eighty miles to the northeast; I-37, taking drivers south to Corpus Christi.

These highways make it easy for people to zip quickly within San Antonio's 417 square miles. But easy freeway access and plenty of space to spread out have encouraged the generic sprawl common to many American cities. It's a place made for cars. The northern part of town is a mosaic of cul-de-sacs, making it hard to get anywhere by walking or biking. The south is crisscrossed with busy one-way thoroughfares with no crosswalks. It's common for people to get into their air-conditioned cars for a two-block trip to the grocery store rather than to dodge a steady stream of cars and to endure the blistering heat.

Driving north or west from the San Antonio International Airport puts you in touch with the city's upper middle class. In this area, simply buying property provides a passport to tennis courts, swimming pools, annual dues, and such august bodies as the Architectural Control Committee, which holds sway over everything outdoors except the shrubbery. One homeowners' association features the Yard of the Month—properly edged, as all must be—and clearly spells out the homeowners' raison d'être: "Community—That's what it's all about in Stoneridge! This is where we live, work, play, cut grass, swim, ride bikes, and keep an eye out for suspicious activities."

Traveling south from there, you encounter other well-to-do neighborhoods with such names as Alamo Heights, Lincoln Heights, and

Terrell Hills. Much of this area shares the 78209 zip code, leading residents to use the term "09ers" as shorthand for the affluent. Further south is downtown, home of the Alamodome where the Spurs play. Here the River Walk along San Antonio River, saved from extinction in 1921 after a devastating flood, is a major tourist draw. At night, with mariachi bands playing and people dining along its banks, the River Walk seems like a magical second city running below the busy downtown streets. Downtown is where much of the city's annual $3.51 billion from tourism changes hands. San Antonio owes much of the tourist trade to its rich Mexican American culture—from Market Square, overflowing with crafts from Mexico's interior, to the many mission-style churches to La Villita National Historic District, an enclave for artists and craftspeople that still has brick-and-tile streets and some of the original eighteenth-century adobe structures.

Steering south away from downtown, you will find some of the country's oldest missions but also a dramatically changed landscape. Head down South St. Mary's Street, and you'll pass Bonham Elementary School, the Texas Highway Patrol Museum, the Adan and Eva Beauty Salon with its hand-lettered sign, and Pig Stands Coffee Shop with a giant smiling pig erected in the parking lot. It's a boulevard of carwashes, taquerias, and aging clapboard buildings. Turn onto any side street, and you'll find rows of small houses, many behind cyclone fences, the windows barred, the gutters cracked, and the paint fading.

In San Antonio, excess pounds and poor health are not confined to any particular group, but it's this part of town where the epidemic of obesity and its twin malady, diabetes, have made the biggest mark. The city's south side has the highest proportion of Hispanic residents (eighty-three percent), its people have the lowest incomes (median family income in 1990 was $19,728 compared to $39,781 in the northwest), and its residents bear the brunt of the city's weight problem. As of 2002, more than seventy-five percent of this area of San Antonio's adult residents were overweight; forty percent were obese. As shocking as the national numbers are, the south side of San Antonio surpassed them long ago.

The Problem with Parents

At the county Women, Infants, and Children (WIC) food assistance program for low-income women and their children, nutritionist Guerrero has a brisk schedule. She counsels mostly Hispanic women whose children are either already overweight or are getting heavy enough to be at risk. Of the 11,000 children in the WIC program, twenty percent are already too heavy. Guerrero, a soft-spoken Mexican American woman with brown eyes and dark brown hair pulled back and swept into a knot, has approximately 150 children in her caseload with weights above the ninety-fifth percentile of BMI. For some of Guerrero's kids, the term "overweight" is woefully inadequate.

One family in particular stands out: an overweight mother in her thirties, her four-year-old who weighs ninety-two pounds, and her other seriously heavy child, age three. Despite prodigious counseling of the mother, the four-year-old has gained two pounds a month for the last nine months. Each month his mother produces a diet recall chart that shows he's not adding calories; in fact, his diet looks pretty good. Guerrero, who offers advice in a manner that's kind but firm, could only conclude that the mother is stretching the truth. The mother has admitted many times that she will not give up eating fried chicken, french fries, and ice cream, and she certainly can't change her children's diet without changing her own. Besides, she's not that worried. Plenty of kids in her extended family are bigger than hers, and there's nothing wrong with them.

A father from a different family was also reluctant to change the family's eating habits, especially when it came to cutting back on sweets in the house. "We don't want to punish ourselves," he told her.

These parents love their kids, but they can't always see the health problem that they're inviting. "They aren't thinking five years down the road," Guerrero says.

Guerrero, WIC Executive Director Melanie Ritsema, and fellow nutritionist Ana Kraft say they face overwhelming obstacles every day trying to encourage a healthy diet and physical activity among the city's poorest families. Poor people in San Antonio—just like poor people in any other U.S. city—have limited food options. In San Antonio, supermarkets on the wealthier side of town devote many more square feet to fruit and vegetable displays than they do in the poorer south side. People who don't have a car are either dependent on public transportation to get to the supermarket, or they're captive to the corner store, where they'll find no fruits or vegetables but plenty of chips, beer, Ho-Hos, and Ding-dongs. "Poverty kind of limits you to where you're at," says Ritsema, a slender redhead who grew up in London. "You don't broaden your horizons in any way, including food. The irony is that in America we have all this food but we're malnourished."

When a woman signs up with WIC, she gets a pack with nutritious foods like tuna and carrots and vouchers for eggs, juice, milk, peanut butter, and cereal. But vouchers don't cover everything, and the women at WIC say they are fighting upstream against a host of cultural norms about weight and health in the Latino community. Many of WIC's clients or their parents came from villages or towns where thinness was equated with poverty and starvation. To them a fat child is a strong and healthy child. Some parents also have an unshakeable belief that their kids will grow into their weight. When researchers at the University of Miami studied low-income Hispanic mothers of overweight children, they found that only twenty-seven percent believed their child's weight was a significant health risk. The women working at WIC have encountered a fatalistic attitude toward weight: everyone else in the family is big. It's no use fighting it.

The biggest reason to fight is type 2 diabetes. In San Antonio, diabetes is now the fourth-leading cause of death, up from the eighth leading cause in 1992. Eleven percent of the city's adults have the disease, almost twice the percentage in the United States as a whole.

San Antonio's health surveys don't include children, but Treviño has some numbers of his own. Of 1,420 elementary school children his program tested in 2001, eighty-nine children, or six percent, had dangerously high blood sugar levels. Extrapolated to the rest of fourth to twelfth graders, he estimates there could be 5,400 children with undiagnosed diabetes in San Antonio.

Clearly, no single culture has cornered the market on unhealthy food. The Alamo City has the same main drag as every other city or suburb, with its lineup of Wendy's, A & W, Burger King, and Chuck E. Cheese. But Mexican food—available on nearly every San Antonio street corner—is the staple cuisine. However, unlike traditional and healthy fare from Mexico, which features rice and beans, fresh vegetables, and small amounts of chicken or meat, the Americanized version of Mexican cuisine includes dollops of cheese, sour cream, guacamole, and fried tortilla chips. And even some traditional recipes are laden with lard. Lard is a key ingredient in tamales and tortillas for tacos and gorditas. A new grandma may be heard to say she intends to turn her newborn grandchild into a "little gordita."

Mexican Americans certainly aren't alone in embracing the idea that food is love. In every tradition, there's someone determined to overfeed the young. But these days fattening kids up is so effortless that the Hispanic grandma—or the Jewish or Italian grandma for that matter—could hardly take pride in the achievement. Fast foods and heavily processed foods are cheap and easily accessible, making them particularly tempting to the poor. If you don't have the money to take your family on vacation or send your kids to summer camp, you can demonstrate your love with a 99-cent bag of Cheetos or a trip to Burger King.

You can show you care about their happiness by personally delivering a McDonald's Happy Meal to your little ones in the school cafeteria. You can knock yourself out by getting up before dawn to make homemade tortillas that will be soaked in oil for enchiladas because this is what makes your kids smile.

Cultural barriers aside, changing the food regime is often at the bottom of a harried mother's to-do list, particularly if she's single. Trained as a registered nurse and lactation consultant, Ritsema used to counsel mothers immediately after the birth of their babies. She would tell them about the advantages of breastfeeding, such as more complete nutrition for the baby and better protection against disease and such maladies as ear infections and diarrhea. Studies even show that breastfeeding protects the baby against obesity later in life. She recalls one woman who gave birth and was heading straight back to work the next day. "She was the only breadwinner. She had nobody else supporting that family," Ritsema says. "She was going back to work before her milk came in, so this baby was going to get a bottle."

Apart from being overwhelmed, the mothers enrolled in WIC don't seem to see the urgency when counselors use the word "overweight," Ritsema notes. "There's an effort always to be politically correct, to say 'overweight,' 'obese,' and things like that," she says. "Sometimes people don't get the message until you mention 'fat.'" Mothers are also more likely to hear the message if it's uttered by a doctor. Guerrero notes that one mother with a 105-pound five-year-old seems unusually receptive to making changes because the pediatrician said her child weighs too much.

Unfortunately, according to a Michigan State University study, doctors rarely address the issue of obesity in children. The study found that between 1997 and 2000 excessive weight was identified in less than one percent of children visiting a doctor's office or urgent care center, even though sixteen percent of the nation's kids are overweight.

The study also found that only seven percent of obese kids actually received a diagnosis of obesity. The doctors surveyed said they didn't have enough time with patients to make the diagnosis. Another report from a survey of pediatricians, pediatric nurse practitioners, and registered dietitians, published in the journal *Pediatrics* in 2002, found that although doctors believed overweight children and teens should be treated, they didn't feel they had the skills to counsel kids effectively or to manage their treatment. Doctors also said their young patients weren't motivated to lose weight and often lacked support from parents as well as other social supports.

Although some of the WIC moms show little interest in nutrition counseling, Guerrero teaches those who are willing how to read nutrition labels, how to determine a reasonable portion size, and the value of fruit and vegetable consumption. "Just the basics," she says, "so they'll be armed with the tools they need to make choices at home." So little is generally understood about nutrition that at WIC nothing is taken for granted. Guerrero said she had one client who was warned by her child's pediatrician that the child was drinking too much one hundred percent juice, so the mom filled her bottle with Kool-Aid instead. Another client, recalls Kraft, was passing up the chance to use her food voucher for healthy cereal and paying out of pocket for the sugar-coated kind. It was more expensive, the woman reasoned, and therefore had to be better for her kids.

Not long ago, the WIC program had to confront its own complicity in the bad food choices that mothers make. Many of San Antonio's WIC sites have vending machines operated by the city. A few years ago they were stocked with nothing but candy, soda, and Twinkies. One day, one of her staff nutritionists burst into Ritsema's office at her limit of frustration. She wanted to know how she could possibly preach good nutrition to her clients "when they're sitting across from me at 8 o'clock in the morning and the kid is drinking Big Red and eating Cheetos, and they say, 'Well, I just bought it out in the lobby.'" At first, the city told Ritsema that mothers simply needed to take more responsibility, as if the moms could keep their hungry kids away from vending machines when they're stuck in the clinic for three hours. As part of San Antonio's Fit City initiative, Ritsema eventually convinced the city and the vendors to restock vending machines with alternatives, labeled "healthier" and "healthiest." (These machines are stamped with the Fit City logo, a rendering of the city's Tower of Americas sprinting along in running shoes. Miller says that although vending machines aren't a critical element in the city's food chain, they serve as a powerful symbol.)

Many major food vendors are on board with the program now, but at the time they thought they'd go broke stocking the machines with peanuts and raisins. Ritsema found this concern absurd. Her typical client has just gotten off a bus in ninety-five-degree heat with her three children. "They're not going anywhere. Their kids are hungry, and she may be pregnant and hungry too," she says. "My point was if you have a captive audience like we do at our clinics, and you fill the machines with healthy foods, that's what they're going to eat because they don't have another option."

Ronald McDonald Takes the Alamo

You haven't really seen the Alamo until you've viewed it through the Golden Arches. In the spring of 2004, McDonald's Corporation unveiled its new adult Happy Meal—a salad, a bottle of water, and a pedometer—at the site of Texas's most famous massacre. Seen through dozens of yellow balloons formed into the famous arches, the historic fortress looked like a Hollywood set that could be rolled up and carted away as soon as Ronald McDonald declared the press conference over. Ronald, second only to Santa Claus in worldwide recognition by kids, was on hand to promote McDonald's "Go Active" campaign: get out there and exercise and then be sure to fuel up with one of the healthy choices at your neighborhood McDonald's.

The whole event was putting the Fit City advocates on edge. How many nice things should they say about McDonald's? Wasn't the company due some praise for trying to do something about their generally unhealthy food? Would supporting McDonald's today come back and bite them tomorrow? Miller—5 foot, 11 inches tall, with short brown hair, an open smile, and a disarming sense of humor—strode to the podium in a purple Walk San Antonio T-shirt and white shorts. She did praise McDonald's Go Active campaign. She also got in a plug for Walk San Antonio, an ambitious program that's enrolled more than 10,000 San Antonians. People register for free at one of eighty-two sites in the city and get a body composition analysis. They exercise on their own, log the amount of physical activity they've had throughout the month, and return to the site monthly for an evaluation of their progress.

Miller thinks that 10,000 participants isn't bad for a city where you take your life in your hands walking across the street. With precious few bike lanes, it's a tough city for bicyclists, too, and Miller's got the scars to prove it. An avid runner and cyclist, she was pedaling along with the green light one day when a car turning right nearly hit her, forcing her to swerve and fall off her bike. The driver appeared so stunned to see a person on a bicycle—and on the ground bleeding from a cut that left an inch-long scar near her right elbow—that he didn't even have the presence of mind to stop. So in a town like San Antonio, if McDonald's wants to promote exercise, she'll take a chance on endorsing them.

Speaking at the press briefing on behalf of the health department, Ritsema, who once picketed a McDonald's in London at the tender age of seventeen, gave a thumbnail sketch of global and local obesity. "There are now more overweight people in the world than there are

hungry people," she said in her distinctive British accent. Ritsema went on to list the health problems associated with obesity—diabetes, heart disease, kidney failure, high cholesterol, high blood pressure, certain cancers—as Ronald McDonald listened politely beneath the podium, occasionally smoothing his magenta hair. Ritsema next described how the county collaborative was working to convince health providers to consider BMI as a fifth vital sign, along with blood pressure, pulse, temperature, and respiration. As for McDonald's she concluded: "We hope this will inspire all restaurants to provide healthy choices and realistic portions for their patrons." That was as far as she'd go.

Olympic gold medal swimmer Josh Davis more than compensated. When the tall, blond San Antonio native stepped up to the mike, he couldn't say enough about his Olympic swim team sponsor. He gushed about how "at home" everyone feels having five McDonald's restaurants in the Olympic Village, how good the free food tastes, how wonderful it is that a kid from San Antonio could rise to become a world-class swimmer. "There's nothing like having that medal draped around your neck," he said as he hung what looked like a plastic facsimile of a gold medal around his neck, "singing that anthem and realizing what a great country we live in.

"I'm so happy McDonald's is providing us with healthier choices so that we can have good energy for our bodies," Davis continued. He ticked off a list of the healthier foods—apple slices, yogurt, the new salads—and added that he, his wife, and their four kids eat at McDonald's all the time, and naturally the kids go for the burgers and fries. "Kids love it, families love it. It makes great memories for our families." He encouraged everyone to "go active" and to keep making family memories at McDonald's.

Earlier that same morning, a second grader at Bellaire Elementary School had made some memories of his own, enjoying an order of hash browns and a stack of McDonald's pancakes dripping with syrup. His mother brought the meal to school as a treat in place of the free, nutritionally approved school breakfast her son would have otherwise eaten. The sweet-faced youngster, at age seven or eight, weighs about 150 pounds and already is showing signs of acanthosis nigricans, a patch of dark skin on the back of his neck that's one of the signs of type 2 diabetes.

People like Josh Davis have no idea, says Gina Castro, edging her car south onto Interstate 35 after the press conference. "The people over there, they don't come to our side of town. They don't suffer from diabetes the way we do. He has no clue."

Diabetes Central

Castro, the administrator who weighed and measured the eight 200-pound fourth graders, speaks from personal experience. Her husband died seven years ago at age forty-nine of a heart attack resulting from diabetes complications, making her a widow at forty-two and the single parent of a five-year-old. Her stepson, who was diagnosed with diabetes at age twelve and is now in his thirties, has already lost a leg to the disease. Castro says her Mexican American husband was very overweight as a child but got thinner and more athletic in adulthood. He was thirty when he was diagnosed with diabetes, but much of the damage was already under way. She's doing her best to keep today's kids in the Harlandale Independent School District from the same fate and has thoroughly indoctrinated her son about good nutrition and exercise. Having a father and other close relatives with diabetes and being part Hispanic puts her son at risk. "He knows how his father died. My twelve-year-old has seen the effects of diabetes," she says. Now he's the kid who walks into a burger joint, orders a grilled chicken sandwich, and chastises his mom for eating fries.

If only it were as easy with Castro's students and their parents. Castro is the CATCH (Coordinated Approach to Childhood Health) program facilitator at the Harlandale district, one of sixteen school districts in San Antonio. Harlandale is a state-recognized school district, one rung below an exemplary district in student performance, even though more than ninety percent of its students are economically disadvantaged. CATCH is a Texas-based elementary school program aimed at improving physical education and school lunch menus while encouraging parent involvement. The program is being used in thirty U.S. states, Canada, and other countries. A study by researchers at Utah State University and the University of Texas at El Paso found that the CATCH program significantly increased moderate to vigorous physical activity during phys ed classes and decreased the fat in school meals. Harlandale, as well as several other San Antonio districts, received CATCH seed money from the health collaborative. Castro, a former physical education teacher, oversees all aspects of the Harlandale program. "I'm the new food police," she quips.

Along with weighing children, with parents' permission Castro tested blood sugar and blood pressure of 1,000 students at four middle schools. She found thirty-eight kids with blood sugars high enough to warrant a diabetes diagnosis. The number of children with high blood pressure was in the hundreds and included most of the kids whose blood sugar was elevated—a one-two punch toward a lifetime of suffering. She was able to inform the parents and suggest they seek medical help but didn't have the staff to do follow-ups. "Our nurses are so overworked, we have so many children on Ritalin and asthma inhalers and other medications, it fell through the cracks," she says.

Change is painstakingly hard, Castro says. Resources are stretched, and parent education is an uphill battle. "My biggest problem," she hesitates to actually say it, "is parents." She remembers a kindergartner who weighed at least 155 pounds. "She came to school wearing a women's size extra large," says Castro. "I talked to the mother. She insisted the child would grow out of it." Castro had to give the five-year-old her own special physical education program. "We didn't make her do things we made the other kids do. We didn't want her to have a heart attack."

The heartbreaking truth is that that little girl wasn't alone. "These children are coming in very obese in kindergarten. By the time we get them they're already 100 pounds at five years old," Castro says, and some of their eating habits are mind-boggling. "The parents have no clue what kind of danger they're causing their children," she says. It's fairly typical these days to see kids carrying soda with them wherever they go, often with the blessing of their parents. Castro recalls a "very obese" family—mother, father, son, and two daughters, the youngest about six years old—who sat in front of her at a recent meeting. All the kids had twenty-ounce soda bottles under their seats. "Like they needed a soda to sit there for a thirty-minute gang awareness meeting," she says.

The blonde, brown-eyed former phys ed teacher has a no-nonsense approach tempered with compassion. She may be the food police, but she's forthright about her own issues. Castro says that she weighed a mere 125 pounds twenty-two years ago when she came to San Antonio from El Paso. Two years ago she hit her high point of more than 200 pounds but has lost much of the weight since then. "I was raised overeating, and I can tell you why my eating habits are horrible," she says. "My parents owned an Italian restaurant; they still do. We had food twenty-four hours a day. My father would not only buy cases of things for the restaurant, he'd buy cases of things for the house. We had food everywhere." Plus her parents were big proponents of finishing everything on your plate. "I don't do that," Castro says. "My son, if he doesn't eat, fine."

With the Harlandale kids who are the most overweight, Castro takes the parents aside and delicately offers eating and exercise advice. Under the CATCH curriculum, it's part of her job. Some parents are appalled she thinks it's any of her business. Others say thank you, what can I do? After one nutrition presentation to a group of parents, in which she cited foods they should avoid, such as Nerds, chips, soda, and other kinds of candy, she had a couple of mothers come up to her afterward in tears. They said, 'That's what I feed my kids."

In other cities, formal notification of a child's overweight status has brought more howls of outrage from parents than appreciation. In 2002, parents in Pennsylvania and Florida accused the schools of meddling and harming their children's self-esteem after receiving letters from their school districts suggesting their kids might have a weight problem. One Pennsylvania mother even appeared on the *Today* show to complain.

The Longest Marathon

On the morning of the Alamo event, Coach Jim Hinkle is warming up the kids in his PE class at Bellaire Elementary School. With all the doors flung open in the non-air-conditioned gym, one hundred second graders are doing warm-up dance steps to a tune that sounds like the house version of the Go-Go's "We Got the Beat." Most of the kids know the steps, but a few are going the wrong way. One makes a wrong turn and ends up on the floor but quickly gets back up on his feet and falls back in line. The point is that they're moving. Looking on, Castro says that aerobic exercise is a key element of the CATCH program. The music shifts to a country swing tune and the kids, most of them smiling, keep going. In this class of one hundred—enormous even by today's standards of consolidating gym classes to save money—about six of the children are truly obese and another eight or nine appear to be overweight. The vast majority are Hispanic. These kids are taking forty-five minutes of phys ed four days a week—an astonishing amount in an era when many schools are cutting back on PE or eliminating it altogether.

Hinkle has white hair and deep blue eyes, and he's wearing a pair of tan shorts and a black T-shirt with yellow letters that say "Bellaire Recognized School." He moves among the kids, confidently calling out direction over the music. They stop a minute and sit on the floor. "Simon says how many of you know about the food groups?" he asks. A dozen hands shoot up. "Simon says how many ate a healthy meal last night?" Eight or nine hands rise. "How many ate great, delicious, greasy potato chips?" A smattering of hands. ("I actually did yesterday," Castro admits, but not within earshot of the kids.) Aside from being the only phys ed teacher at Bellaire, he's also the only health teacher, and PE is where the students get their nutrition education. Hinkle's two assistants then circulate among the crowd handing out jump ropes and hula hoops.

With the music blaring and his two assistants in charge, Hinkle takes a minute to talk about how his job has changed in the twenty-five years since he came to Bellaire school. For his first fourteen years, every student regardless of age ran a mile every day. "I worked the hell out of them," he says. "If I took all 600 kids, every one of them would pass the fitness test, including the special eds. If we give the 600 today the same test, maybe fifty could pass." But once his buff elementary school kids got into middle school, where phys ed was more lax, their fitness level plummeted. "As time went on—I live in this area, and I'd see them—they would be ballooned out of shape. They weren't educated [about health and fitness]. They were physically fit, but as far as that education in the mind, no." Today's kids may be fatter and in poorer physical shape in elementary school, but they're better educated about making healthy choices, and because phys ed is more fun, the hope is that this generation will find exercise they like and stick with it.

In a sign of the new PE's power to motivate, this year every one of Hinkle's students successfully completed a yearlong marathon. Kids could walk the twenty-six miles in increments as small as a quarter mile. For the young ones who couldn't go walking alone, it meant that other family members had to get out and exercise with them.

By now the kids are sweating, and they're taking turns running half a lap on the grass outside and running back in. The boy who had the McDonald's breakfast is clearly the largest of the group at about 150 pounds. He's wearing shorts and a T-shirt with blue stripes. His black hair is in a crew cut, and he has a large pudgy face, thick legs, and a round belly. He's unable to run around the cones outside but trudges slowly past them instead, the last kid in his group to come in. Asked how parents respond when they're warned that fast food for breakfast might be a bad idea, Hinkle shrugs. "We don't talk to the parents about that. They'll be running to the office [to complain]," he says.

But because he's been in the district so long and taught so many of the kids who are now parents, Hinkle has a pretty good rapport with most of them. He tries to take the kids' height and weight measurements every nine weeks, and if a child is overweight, he will tell the parents. Some are offended, others are grateful.

Partying with Bananas

Among the many ironies in this age of obesity among schoolchildren, none stands out like the sale of junk food to finance physical education. Across the country, PTAs hold bake sales to resurface the track, or the schools themselves hawk candy, soda, and chips to students so that next year they can buy a few stationary bicycles. Those practices are starting to be viewed with horror by many parents, and they're now verboten in some school districts, including Harlandale. Having banned candy as a fundraiser, Hinkle sells the kids *paletas*, Spanish for Popsicle, that are made from one hundred percent fruit juice. On this day, though, Hinkle opens up the freezer to find a vendor has mistakenly stocked it with fruit bars that have corn syrup. This rankles Castro, who makes a mental note to chastise the appropriate person. Hinkle can make $1,000 profit in six months from selling the frozen treats. Then he'll plow the profits into buying more equipment.

The *paletas* were second-grade teacher Kaye Lucas's idea. Lucas, who also serves as PTA president, decided last year to ban sweet and fatty treats from her classroom. Immediately following Coach Hinkle's PE class, her students had a send-off for their student teacher, and party treats consisted of half a banana each, which the kids were gobbling up hungrily. Lucas says that in the past, her kids would have celebrated with cupcakes or candy. Now when they bring in a snack for the whole class, it's string cheese, raisins, or peanuts. She says that these days her kids are calmer in class, and she worries less about snack time undermining their health. The boy who couldn't run a lap said he loved bananas as he munched on his. Bananas do have their drawbacks, however. The boy next to him kept trying to get Lucas's attention. "Miss," he said, shoving his snack toward his teacher, "I can't open this."

Schools seeking to kick their junk food habit might well look to the state of Texas for guidance, where fatty foods were cut way back and junk food and soda sales were severely restricted beginning August 1, 2004. Much of this has been under way in Harlandale for years. For about the last decade, under Sally Cody's direction, junk food sales have taken a nosedive. Cody, the Harlandale district's food service coordinator, has been excising certain foods because of their fat or sugar content—the packaged foods such as cookies and chips as well as the meals in the government-regulated school lunch and breakfast programs. Her budget has taken a hit for it too because, not surprisingly, the highest-fat items are the most popular with kids. Take Flaming Hot Cheetos, for instance, a snack food that appears to be an enticement to kids in every corner of the nation. "They're really good," says Cody. "You have to have a glass of cold water with them, though. Until your system adjusts, you're on fire. But they're so good." At twelve grams of fat per serving, they are not so good for you, though. Cody must not have been alone in banning the snack because Frito Lay recently came up with a four-fat-gram version designed specially for schools.

Ice cream was also a big seller but not anymore. Astonishingly, Harlandale High School used to sell full quarts of ice cream to students. "I don't even want to know how much fat is in a quart," Cody says. "It's a big, big seller, but we had to take them out. The dilemma of a parent is that you give your kids money; you don't know what they're buying. So what we have to do on our end is be careful what we're offering if we want to be responsible for promoting their good health."

Under a federal plan called Provision 2, school districts in Texas with primarily poor families must provide free school lunches and breakfasts regardless of a family's ability to pay. Although Cody supports the plan because it guarantees that all kids get a meal, it puts her in the red financially before she even opens the cafeteria doors. It means she loses more than $400,000 a year in lunch money from kids who could pay either the full price or something less. She makes a portion of that up in government reimbursements every time a kid picks up a school lunch, but still, it's a losing proposition. Nonetheless, some things are better than they used to be. Cody used to get penalized for banning butter and fat from her cafeterias. The USDA, while publicly promoting less fat in meals, used to send her "mountains" of shortening and butter, and if she didn't use up her government commodities, she'd get fined.

For at least as far back as three generations in San Antonio, Wednesday is enchilada day—whether you grew up in a Hispanic family or not. But enchiladas have a huge amount of fat. Ten years ago, Cody set out to remodel this iconic food but not without a bit of resistance from her cafeteria managers, who above all else, know how to make an enchilada. First, she had them stop dipping the tortillas in oil; they were steamed instead to keep them soft. As for side dishes, she made sure there was no fat added to the rice and no lard or shortening in the beans. The only ingredient with fat is the cheese. "They've made it taste wonderful," says Cody.

Today, there's not a single deep-fat fryer left in the school district. Cody is actively promoting fruits and vegetables but also walking that fine line between what's healthy and what kids will eat, while maintaining a reasonable cash flow. Each school serves a baked potato and a chef's salad every day. There's also a daily choice of fresh or canned fruit and a growing assortment of fresh vegetables—baby carrots, raw cauliflower, broccoli with low-fat ranch dressing, and *calbacita*, a squash and corn mixture with a spicy back note.

Cody is hoping that over time, the kids will get used to the kind of healthy eating that will carry into high school. She would prefer that parents get used to it too, especially the ones who keep bringing fast food into the lunchroom. The healthier foods are catching on slowly with the kids, but in truth it's been a tough sell, especially with the younger ones. Veteran cafeteria manager Carmen Martinez, who works for Cody, says children used to eat better when she started twenty-five years ago. Now all they want is pizza and fries, she says with more than a hint of sadness.

Moving Mountains

By his own account, Roberto Treviño used to be a hoodlum, running the streets of San Antonio's housing projects. Of Mexican heritage, the father of two teenagers has come a long way since then, but not geographically. Treviño works no farther than a few blocks over from where he grew up. His parents, whom he credits with helping him realize his potential, still live across the street from their former housing project. A successful internist who opened three practices in the old neighborhood, he could still be earning $280,000 a year and hewing to his roots—but for a nagging feeling that he wasn't answering his true calling.

For thirteen years Treviño did the best he could with his adult patients, tending to their aches and pains and general maladies. But it was the advanced type 2 diabetes, rampant in his community, that finally inspired him to make a change. He was tired of referring people for kidney dialysis and watching the disease's complications ravage his patients' bodies. "It's too late in the assembly line when they come into my office with blood sugars of 300," he says. The research community offered little help. As a researcher and clinician, Treviño himself had been widely published, but neither his nor anyone else's findings were making anyone healthier. "You go around the country and all you hear about is 'health disparities and minorities' and 'diabetes is getting worse,' but nobody's doing something about it," Treviño says.

Where to start? Treviño reasoned that if he was going to deal with prevention, adulthood was too late; he'd have to start with children. And his program would not change just knowledge (knowledge isn't stopping people from smoking) or beliefs (everyone knows you are what you eat) but would also target behavior. He set up a nonprofit called Bienestar ("well-being" in Spanish) in a wood-frame house with another building out back, a few blocks down from Pig Stands Coffee Shop and around the corner from Taco Haven. He would see if he could stop the march toward diabetes.

Treviño has short, thinning black hair flecked with gray and the kind of frenetic energy it takes to work twelve-hour days between his nonprofit and his medical office, where he still keeps limited hours. At his Bienestar office, he's surrounded by old family photos that look like they might belong in one of San Antonio's history museums. In one, a handsome mustached man astride a white horse stares soulfully into the camera. The man is his grandfather, Antonio Treviño, who later died in the Mexican Revolution. Antonio's grandson Roberto has the same sort of revolutionary fervor when it comes to his community. "This is not a job," he says. "This is a mission."

Treviño's program is based on social cognitive theory, which is essentially the idea that creating social support can lead to behavioral change, and it's a passion of his. "If you go to work and your coworkers are telling you, hey, you've got to eat more fruit, and you go home and your family's telling you, hey, we need to go out and walk on weekends, and you go out with your friends and your friends are telling you, hey, you've got to eat more vegetables, that peer pressure, that social pressure is more likely to mold your behavior," he says. "That's where this program originates—that we need to change the social systems that influence a child's behavior."

Treviño started Bienestar in 1994 as a school-based diabetes prevention program targeting physical education, school food, family, and after-school programs. In 1999, with the help of a $2 million grant from the National Institutes of Health (NIH), he began a three-year randomized clinical trial with one set of kids receiving intervention (the kids would call it having fun after school) and a group of children serving as controls. The vast majority were very low-income Mexican American kids. The staff devised workbooks for parents, PE teachers, and school food service personnel with the objective of decreasing the children's dietary fat, increasing fiber intake, and raising their level of physical activity—all behaviors known to decrease body fat and lower blood sugar levels. Bienestar also held an after-school program once a week where the kids exercised and, along with their parents, received cooking and nutrition lessons.

Over each of the thirty-two-week trials, blood sugar decreased in children with abnormally high levels. The Bienestar program brought down dangerously high blood sugar in ninety-seven children. The children in the intervention group also increased their fitness levels and their fiber intake. Body fat decreased but not significantly compared to the control group. Treviño's theory is that at age nine it's almost too late for that. For one thing, the body creates most of the fat cells it's going to have for life between age four and age nine. Those cells are permanent, even though they can get bigger or smaller. Also, all children, even average-weight kids, undergo something resembling a diabetes state around the time of puberty between the ages of nine and thirteen—that is, their insulin levels, blood sugar, and body fat increase.

Even though these levels increase in average-weight kids, they stay within normal ranges. But if a child is already overweight, the changes could signal impending diabetes. Around that age it's not only their biology that changes; their behavior shifts in an unhealthy direction too. Studies show that low-income children begin to exercise less and eat fewer fruits and vegetables and consume more sugar. Treviño's answer is to start interventions in kindergarten, next on his list of plans after he launches a study of middle school kids with a new $4 million NIH grant. "When we do kindergarten through middle school, I think it's going to be awesome," he says.

Treviño notes with considerable pride that his 32-week studies are longer than the usual 26-week trial the pharmaceutical companies run before they bring a drug to market. And his solution is drug-free and costs a lot less—about $190 a year, compared to $1,200 a year for one diabetes drug.

Some of the benefits, when his families really take the message to heart, are priceless. Alma Lopez has been on a healthier eating quest ever since Bienestar health educator Janie Centeno taught the parents how to make a fat-free yogurt and granola parfait. She showed them a few other tricks too, such as how to prepare an Oreo smoothie with fat-free ice cream, reduced-fat Oreos, and skim milk. An instant convert, Alma and her ten-year-old daughter Ameri left the after-school program and headed straight for a few aisles in the grocery store they had never even noticed before. "I'm Mexican, and you know you cook with a lot of oil, you eat a lot of tortillas, *manteca* (lard), the real fattening things," Alma says. "I try to cut down on the fat because my family is diabetic on both sides." Alma, forty-one, has a long family history of diabetes and had gestational diabetes herself when she was pregnant with Ameri and Ameri's older brother.

Ameri, a strikingly pretty fourth grader who wears her dark curly hair in a ponytail, doesn't look particularly overweight, but her ethnic background and family history put her at risk for both obesity and diabetes. Regardless of weight, Ameri has learned some valuable lessons that would reap benefits for any kid. "I like when I learn that I shouldn't watch TV too much, and I shouldn't eat too much junk food, and I should exercise every day," Ameri says. She says now she only watches TV for thirty minutes a day compared to before when she watched one hour daily. Her mother smiles and holds up two fingers: correction, two hours a day. Her mother says that left to her own devices, Ameri could have easily spent three or four hours a day in front of the tube. The family recently moved, and her mom says they have such a big backyard that these days Ameri just watches one show, then runs out to play.

"Now I like to play kickball, and I also like to play soccer with my friends and volleyball with my dad, and I exercise with my dog too." Ameri says she has more energy now than when she joined the program four months ago, and she doesn't put up a fuss if after the family walk, her mom serves up yogurt and granola or carrots instead of potato chips. Ameri is eating fewer chips and more fruit although she still loves Flaming Hot Cheetos, and so do her friends. "One of my friends will even lick the bag," she says. These days, Ameri warns her friends that eating a lot of Cheetos is unhealthy and could make them sick.

The program so inspired her mom that next she wants to take a low-fat cooking class offered for free through the Texas Diabetic Institute. Recently, Alma had Ameri's blood sugar tested because, knowing what the symptoms are, the youngster was worried she had some of them. Ameri's blood sugar was normal, and with her family's help, it should stay that way.

As for Treviño, he's rolling up his sleeves for the next round of clinical trials and grateful for the view from his second-story window at Bienestar. At his medical practice, he says, "I look out the window and wonder what I'm doing. Over here, I'm having a great time. These very low-income children, they see you, and they hug you. You'll move mountains for these kids. So you keep going."

3

Babes in Calorie Land

THE GOLDEN CORRAL, now open in forty-nine states (sorry, Hawaii), is more than just an all-you-can-eat buffet restaurant: it's a window into the dark desires of children. Little boys and girls stand on their tiptoes, peer into the steaming trays of food, and ask mom or dad for fried chicken, pepperoni pizza, and macaroni and cheese. The big kids rush from one station to another, loading their own plates with, as it happens, fried chicken, pepperoni pizza, and macaroni and cheese. Some kids pile on enough food to threaten the structural integrity of their plates, but they all manage to leave room for a trip to the dessert bar, the place where their depravity takes full bloom. They pour gummy bears on fudge brownies. They sprinkle candy corn on top of vanilla soft-serve ice cream cones. And they go back for seconds.

Today's kids are savvy about nutrition. Most elementary school children can rattle off the six food groups, and many teenagers can recite the USDA recommendations for daily servings of fruits and vegetables. But let them loose in the Golden Corral, and the truth comes out: kids are perfectly happy making horrible choices. If it were up to them, the whole world would be a dessert bar.

There was a time when the American diet had a distinctly adult flavor, even when it was served to kids. A cookbook published in 1966 offers the following dinner menu for the preschool child: liver and potato pie, peas cooked in lettuce, minced uncooked cabbage with lemon juice, baked custard, milk. Looking back, that meal sounds about as modern as a roasted hog's head. Today's kids expect more "fun" in their meals, and they get it.

Restaurants and food manufacturers are constantly looking for ways to make foods more appealing to kids. They put goofy characters on the labels, they give away toys, and most of all, they steer clear of things that might be offensive, such as fruits, vegetables, or nutrients. The nutrition professor Marion Nestle eloquently summed up the situation at the 2004 Summit on Obesity in Williamsburg, Virginia: "These companies are in business. They are not sitting around a table saying, 'Let's see how we can make kids fat.' They are saying, 'How can we sell our product in a marketplace that is extremely competitive and in which there is too much food around?' And they're trying to do three things: they're trying to establish brand loyalty as early in life as possible; they're trying to get kids to pester their parents to buy more food; and the most insidious of all is that they're trying to get kids to think that they're supposed to have their own special foods, like Lunchables and other things in packages. So kids are not supposed to eat the boring foods that their parents eat."

In recent years, the entire American diet has taken a decidedly childish turn. As a nation, we're getting less food from the family dinner table and more from fast food cartons, snack bags, and candy wrappers. Kids should be thrilled—things could hardly be better if they were in charge—but there's a downside to their triumph: they're getting fatter.

Feeding kids well has always been a struggle. A hundred years ago, many children suffered from rickets because of a shortage of vitamin D. Others developed cretinism (a form of mental retardation) because they weren't getting enough iodine. Far more children simply couldn't get enough calories to support their active, hardworking bodies. Malnourishment left them spindly, sickly, and short.

After a hundred years of progress, we've moved from rickets, cretinism, and malnourishment to obesity, the nutritional disease of a new age. Unlike rickets or cretinism, obesity has no single cause, and solving it won't be as simple as putting vitamin D in milk or iodine in salt.

One Big Happy Meal

The epidemic of childhood obesity really started picking up steam in the mid-1970s, and, not coincidentally, that's also when the American diet started regressing to more childish fare. Perhaps the clearest look at dietary changes comes from comprehensive surveys of more than 60,000 people conducted between 1977 and 1996. For all ages, consumption of salty snacks more than doubled in those nineteen years. The percentage of calories coming from pizza, french fries, candy, and soda also rose sharply across the board.

The surveys didn't specifically ask about fudge brownies covered with gummy bears, but the trend isn't hard to guess. As a nation, we have developed a sweet tooth that just keeps getting sweeter. Americans added about eighty-three calories' worth of sweeteners a day to their diet (not including sugar naturally found in foods) during those years. Soft drinks accounted for fifty-four of those extra calories.

Our diets are getting healthier in some ways, at least by the loosest definitions. The percentage of calories that come from fat actually seemed to decline slightly between the early 1970s and 2000, but our diet is still too fatty. According to the CDC, Americans of all ages got about thirty-three percent (or one third) of calories from fat in 2000, which is more than the recommended limit of thirty percent. Grain consumption increased between the mid-1970s and the mid-1990s, largely thanks to breakfast cereal, corn chips, pretzels, popcorn, and crackers. Still, less than one-fourth of people got the recommended daily servings, which range from six a day for young children to eleven for active men. (With the recent low-carb craze, the number of people who reach these standards has undoubtedly dropped.) Americans started eating more fruits and vegetables in the 1990s, but thirty percent of vegetables were potatoes, including a hefty dose of french fries.

There's no doubt that kids are at the leading edge of America's diet revolution. They are the innovators, the risk takers, the only ones willing to try new Twisted Cheetos. (The slightly ambiguous television ad for this product proclaims "They'll turn your tongue green . . . for a limited time.") Whether they're in kindergarten or high school, kids are seeking out more and more treats and leaving healthier food behind. By the mid-1990s, kids were getting a full one fourth of their calories from desserts and sweeteners. Between the late-1970s and the mid-1990s, soft drink consumption increased by sixty-five percent in adolescent girls and by seventy-four percent in adolescent boys. Boys now average about nineteen ounces of soda a day, or more than one and a half cans' worth. Meanwhile, older kids and teens cut back on milk by thirty-six percent between the 1960s and the 1990s. These shifts left most kids miles away from standard dietary guidelines and are helping to set them up for weak bones and osteoporosis later in life. According to a 2003 report in the *American Journal of Clinical Nutrition*, all signs suggest that only one percent of kids—at most—meet all of the recommendations of the USDA food pyramid for children.

Many experts now see a direct link between the radical shift in kids' diets and the explosion of childhood obesity. As pediatrician Arnold Slyper points out in an article published in 2004 in the *Journal of Clinical Endocrinology and Metabolism*, all of those chips, buns, fries, and sodas have something in common: They are all loaded with simple carbohydrates, little packages of easily digestible energy that can quickly turn to fat. At the same time, kids are eating fewer and fewer foods rich in complex carbohydrates, such as whole grains, fruits, and vegetables. Slyper speculates that the volume and the type of carbs that kids eat "may be the most important contributors to our pediatric obesity epidemic."

To understand fully how far children have strayed from dietary ideals, it helps to know what those ideals are. (Table 3.1, on the following page, provides a summary of the latest wisdom on childhood nutrition. Chapter 7 contains a more complete discussion on feeding kids at every age.)

Nutritional guidelines were the last thing on Tracy Graham's mind as she fed her family. The Clinton, Connecticut, mother of three worked sixty hours a week as a nuclear medicine technician at a local hospital, and she had just a few basic standards for the family meal: it had to be fast, it had to be cheap, and it had to be something her kids would actually eat. McDonald's became a second home, or at least a second dining room. On nights when she worked late, she would leave the kids her credit card number so they could order a pizza or two. When she did cook, she tried to make enough to last for a week, but it was often gone in a couple of days.

Her two oldest kids, a boy and a girl from a previous marriage, have always been tall and beanpole thin, so she never really worried about how much they were eating. But then her youngest son Josh came along. He was chubby in grade school, overweight in middle school, and huge by high school. Even as he went past 300 pounds and then 400 pounds, the family still made their trips to McDonald's, where they supersized as always. "We knew Josh was overweight, but we didn't know how to make healthier choices," Tracy says.

A busy mom, a hungry family, yet another meal on the go—the story of the Grahams would be familiar to countless Americans. For all of the talk about fats, carbs, and calories, the biggest change in the American diet has been the move away from the home-cooked meal. According to a 2002 report by USDA researchers, in 1996 Americans got about one third of their daily calories from foods prepared outside of the house, a nearly one hundred percent increase since 1977. In 1999

Table 3.1 Feeding Guidelines for Children

		Food Groups			
Age	Breast Milk or Infant Formula	Bread and Grains	Fruits	Vegetables	Lean Proteins
Birth to 3 months	20-24 oz (6-10 feedings/day)	None	None	None	None
4-5 months	24-32 oz (4-6 feedings/day)	Offer baby cereal pre-pared with water, breast milk, or fomula (3-4 Tbsp/day)	None	None	None
6-7 months	24-32 or (4-5 feedings/day	4 or more Tbsp of iron-fortified baby cereal	Offer cooked, soft or mashed baby fruits and vegetables. Slowly advance to 4-5 Tbsp/day. No more than 3-4 ounces of fruit juice		None
8-9 months	24-32 oz (3-4 feedings/day	4 or more Tbsp of iron-fortified baby cereal	4 or mor Tbsp	4 or more Tbsp	Offer cooked, soft, minced or pureed meats, poultry, cheese cubes, egg yolk (avoid egg whites, shellfish, nuts, and peanut butter until after age 1).
10-12 months	24-32 oz (3-4 feedings/day)	1/4 c of iron-fortified baby cereal, other soft, cooked starches	1/2 - 3/4 cup	1/4 - 1/2 cup	2-4 Tbsp

		Food Groups			
Age	Breads and Grains	Fruits	Vegetables	Dairy Products	Lean Meats, Fish, Poultry, Beans, Nuts, and Eggs
1-3 years	6 servings 1/2 slice bread 1/4 - `/3 c cooked cereal 1/4 - 1/3 c cooked rice, pasta, noodles 1/4 - 1/2 c dry cereal	2 servings 1/4 - 1/2 pc small fruit 1/4 - 1/3 c fruit juice* 2-3 Tbsp canned fruit *Limit to 4-6 oz/day	3 servings 2-3 Tbsp cooked or finely chopped raw 1/4 medium potato	2 servings 1/2 c milk 1/2 c yogurt 1/2 oz cheese	2 servings 1/2 - 1 oz cooked lean meat 1/3 c cooked beans 1/2 - 1 egg
4-6 years	6 servings 1 slice bread 1/2 - 3/4 c dry cereal 1/3-1/2 c cooked cereal 1/3 - 1/2 c cooked rice or pasta	2 servings 1/2 - 1 piece small fresh fruit 1/2 c fruit juice* 1/3 c canned fruit *Limit to 4-6 oz/day	3 servings 1/3 c chopped raw or cooked	2 servings 3/4 c low fat milk 3/4 c yogurt 1 oz cheese	2 servings 1 1/2 oz cooked lean meat 1/2 c cooked beans 1 egg 1-2 Tbsp peanut butter
7 and older	6-11 servings 1 slice bread 3/4 - 1 c dry cereal 1/2 c cooked cereal 1/2 c cooked rice or pasta	2-4 servings 1 medium piece fresh fruit 3/4 c fruit juice* 1/2 c canned fruit 1/4 c dried fruit *Limit to 8-12 oz/day	3-5 servings 1/2 c raw or cooked 1 c leafy	3-5 servings 1 cup low fat milk 1 c yogurt 1 1/2 oz cheese	2-3 servings 2-3 oz cooked lean meat 1/2 - 1 c cooked beans 1-2 eggs 2-3 Tbsp peanut butter

Americans spent about half of their food money on snacks and meals prepared away from home, and that doesn't even count all of those instant soups or microwavable TV dinners that don't require any real cooking.

Table 3.2 Cost of Fast Food vs. Home-Cooked Meal

Fast Food Meal for 4		Home Cooked Meal for 4	
Big Mac, Large Fried, Large Soda	$5.58	Oven Stuffer Roaster	$5.99
McGrill Chicken Sandwich, Large Fries, Large Soda	$5.98	Baked Potatoes	$2.99
Happy Meal with Chicken McNuggets	$3.39	Broccoli	$1.78
Quarter Pounder with Cheese, Medium Fries, Medium Soda	$5.09	Tub of Margarine	$2.39
2 McFlurry's	$7.18	1 Gallon Low Fat Milk	$3.89
2 Cones	$2.18	Watermelon	$1.29
Total	**$29.40**	**Total**	**$18.33**

"Food away from home has become an affordable treat," says Joanne Guthrie, PhD, a nutritionist and lead author of the USDA report. In many families it has also become a near necessity, with huge numbers of women working outside the home. Kids, too, have been getting a lot more of their food outside of the house, which is the best news Ronald McDonald could ever hope for. The percentage of kids' daily calories that come from fast food increased fivefold between the late-1970s and the mid-1990s. On any given day, about thirty percent of kids will eat fast food, and many eat it every day. With roughly 250,000 fast food restaurants in this country—including outlets in malls, hospitals, and gas stations—kids don't have to go long between their french fry fixes.

Low prices are definitely part of the allure of fast food. An entire meal at Burger King can cost less than a single appetizer at Applebee's. But when compared with home-cooked meals, fast food can be very expensive. For the price of a Whopper at Burger King, you could make a hamburger at home . . . with two pounds of beef. Not that you'd want to. An eight-piece bucket of fried chicken from KFC costs a little over $9 dollars. For comparison, six servings of oven-fried chicken—a whole broiler chicken baked with a coating mix plus a little paprika and salt—costs a little over $4. (See Table 3.2 for a another cost comparison.)

For real bargains, you don't have to look further than the produce aisle. A nationwide survey conducted by the USDA in 1999 identified more than forty fruits and vegetables that cost less than fifty cents per serving. The survey also found that careful shoppers could get their recommended four servings of vegetables and three servings of fruit for as little as sixty-four cents a day.

Homemade meals are also less expensive than heat-and-eat, store-bought meals, at least when you consider the cost per serving. But in another sense, store-bought convenience foods are a sort of "bargain." Frozen dinners and other ready-to-eat foods often pack a remarkable number of calories. Marie Callender's Chunky Noodle and Tuna entrée has 960 calories, or nearly half what the average person needs in an entire day. It's more expensive than any tuna and pasta dish you would whip up at home, but it has more calories. As a result, you pay more *per calorie* (not per dish) when you make food at home.

Or consider a new "kid-friendly" offering from Stouffer's: Maxaroni (macaroni and cheese) with a side of cheese pizza rolls. An eight-ounce serving contains 430 calories and twenty-two grams of fat, and it can all be yours for about $2.30. You'd almost certainly pay more for that many calories if you made the same dish at home. For the sake of argument, we'll overlook the fact that few people would ever combine cheese pizza with macaroni and cheese at home.

When noted food researcher Adam Drewnowski, PhD, of the University of Washington surveyed about 200 products at a Seattle supermarket, he found that the individual calories in convenience foods usually cost less than the calories in fruits, vegetables, meat, and other unprocessed foods. In other words, families who rely on frozen dinners and other ready-to-eat meals are getting plenty of bang—and quite possibly bulge—for their buck.

Table 3.3 Children's Calorie Needs

Age	Daily Calorie Needs
0-6 months	650
6 months-1	850
1-3	1,300
4-6	1,800
7-10	2,000
Males	
11-14	2,500
15-18	3,000
19-24	2,900
Females	
11-14	2,200
15-18	2,200
19-24	2,200

Calories with a Side of Calories

Families save time and dish soap (if not money) when they go out to eat, but they lose something extremely important—control. Even the most ardent label readers and calorie watchers tend to run into trouble when they walk into a restaurant, Guthrie says. For one thing, they can no longer choose the portion sizes or the ingredients. And, like kids at a dessert bar, they often lose their inhibitions when tempting foods are laid out before them. "People have different rules about what they eat at home and what they eat out," Guthrie says. Even if they walk in planning to get a salad, the smell of french fries or fried chicken can overwhelm their best intentions, she says.

If it's hard for adults to eat sensibly outside the house, it can be practically impossible for children, especially if they order from the children's menu. As the Center for Science in the Public Interest (CSPI) reported in March 2004, the children's menus of most major chain restaurants are a nutritional disaster. Just about every option is deep fried or greasy and, aside from the occasional pickle, completely lacking in greens. Kids ordering from the children's menu at Applebee's have a choice of chicken fingers (with fries), a cheeseburger (with fries), a grilled cheese sandwich (with fries), a corn dog (with fries), and macaroni and cheese (fries available upon request). According to CSPI, that grilled cheese sandwich with fries has 900 calories and more saturated and trans fats than a kid should eat in an entire day. (See Table 3.3 for breakdown of daily calorie needs.)

Of course, the kid's menus at fast food restaurants aren't any healthier. One of the aptly named "Big Kids" meals at Burger King includes a double cheeseburger, small fries, and a sixteen-ounce Coke. That's 920 calories and forty-two grams of fat in one meal.

Whether they go to Burger King, McDonald's, Taco Bell, or another fast food restaurant, kids tend to walk away full—and then some. A study by researchers at the USDA and Harvard University in 2004 found that kids get 187 extra calories on days when they eat fast food. While newspapers greeted the study with alarming headlines, defenders of fast food responded with ridicule. The Center for Consumer Freedom, a food industry front group, noted on its website that 187 calories "is equivalent to less than a cup of lima beans . . . Horrors!"

But even a few extra calories can really add up over time, says Shanthy Bowman, PhD, a nutritionist with the USDA and the lead author of the study. "It takes about 3,400 extra calories to gain a pound," she says. At 187 extra calories per visit, a kid would have to eat fast food about thirty to thirty-five times to put on a pound of fat. "If children have fast food once a month, it won't be a problem," she says. "But if they're eating it three times a week, they're consuming more energy than they need." Although fast food is almost certainly helping many kids put on weight, it's hardly the sole culprit, Bowman says. "I would never say that fast food is the only reason why people are overweight," she says. "I go to KFC three times a year, and I still gain weight."

Kids who are already overweight have every reason to go easy on fast food, but they may have a particularly hard time fighting the temptation. Harvard Medical School researcher Cara Ebbeling, PhD, recently took a group of teenagers to a local food court and let them eat their fill. The heavy kids quickly put away about 1,800 calories' worth of chicken nuggets, fries, cookies, and cola, but the thin kids stopped at about 1,400 calories. Ebbeling concluded overweight children have "some sort of susceptibility" to fast food, but she hesitates to even speculate about why that might be. Suffice it to say that supersized kids tend to bring supersized appetites when they eat out, a fact that could easily lead to even more unwanted pounds.

Children of all sizes are also likely to overeat when given outrageously large portions, says Jennifer Fisher, an assistant professor of pediatrics at Baylor College of Medicine in Houston. In a study published in 2003, Fisher and colleagues gave thirty preschoolers child-sized portions of macaroni and cheese for lunch. The kids ate their fill and were ready to go play outside. On the next day, each child was again offered macaroni and cheese, but twice as much as before. When faced with a mound of food, kids ate twenty-five percent more than they did before.

Out in the real world, child-sized portions are practically disappearing. Remember when muffins used to be a light breakfast snack? If you buy one at a coffee shop today, be sure to lift with your legs, not your back. According to USDA surveys, portion sizes for salty snacks, soft drinks, french fries, cheeseburgers, and Mexican food all increased dramatically between 1977 and 1998. The size of the typical cheeseburger, for instance, went from 5.8 ounces to 7.3 ounces, a jump that added 136 extra calories. Even toddlers are supersizing: between the 1970s and the 1990s, children ages one year to eighteen months were given increasingly large servings of milk, bread, cereal, juices, and peanut butter, according to the surveys.

Portion inflation is especially severe away from the home. "Look through the window of any of the big chain restaurants, and you'll see huge platters of food coming out of the kitchen," says Melanie Polk, registered dietitian and director of nutrition education for the American Institute of Cancer Research. "It's called value marketing," she says. As the food industry grows more and more competitive, restaurants begin to lure customers with bigger and bigger meals. "They are always offering more for less. It would please me if they started offering less for less."

"All of the Beverage Opportunities They Need"

Value marketing has also inflated the size of the typical serving of soda. The twenty-ounce bottle has largely replaced the twelve-ounce can, especially in vending machines and convenience stores. That's about 250 calories and seventeen teaspoons of sugar that can disappear in a couple of minutes on a hot day. But there's more behind the ongoing soda craze than bigger bottles. Back in the early 1970s, manufacturers started replacing the sucrose (sugar) in sodas with high fructose corn syrup, a makeover that made soda less expensive and more attractive to young consumers. "Fructose is much sweeter and gives more of a bang for the buck," says George Bray, a professor of medicine at Louisiana State University. "Since sweet taste is a preferred taste—in nature it was usually associated with other healthful things—more sweetness drives more intake."

In a 2004 commentary in the *American Journal of Clinical Nutrition*, Bray noted that high fructose corn syrup has other qualities that might encourage kids (and adults) to go overboard. Unlike other types of sugar, corn syrup doesn't trigger the release of leptin, a hormone that tells the body when it's full. That partly explains why a kid can chug a quick 250 calories and still thirst for more.

With all of those calories in one convenient package, it's no surprise that some kids manage to guzzle their way to a weight problem. The mother of one nine-year-old girl in the Committed to Kids weight loss program at Louisiana State University couldn't understand why her daughter weighed more than 300 pounds. After all, the girl never seemed to eat very much. When questioned by the staff, the daughter said she drank four or five twenty-ounce sodas—or 1,000 to 1,250 calories—a day. Another mystery solved.

This girl was an extreme case. Few kids can stomach 100 ounces of soda in a day, and even fewer ever reach a weight as high as 300 pounds. But plenty of children drink way too much soda, and there's strong evidence that all of those liquid calories can turn to fat. In a first-of-its-kind study, Harvard researchers tracked both the weights and the soft-drink consumption of 548 grade school children for two years. As reported in the February 17, 2001, issue of *The Lancet*, each daily serving of a sugary drink seemed to raise the risk of obesity by sixty percent.

The *Lancet* study sounded alarm bells among nutrition experts and sparked an avalanche of news reports. It also spurred proponents of soft drinks to action. The National Soft Drink Association and similar groups started campaigns to restore the image of soda. Even the highly respected American Dietetic Association (ADA) published a fact sheet—"Straight Facts about Beverage Choices"—that cast soft drinks in an essentially positive light. The fact sheet starts with the assertion that "all beverages can have a place in a well-balanced eating pattern." About seventy-five percent of the text is devoted to dispelling fears about soda.

Why would the ADA take the side of soft drinks? Are our nation's nutritionists worried that kids aren't getting enough sugar? Hardly. The fine print tells the real story: "Straight Facts about Beverage Choices" was sponsored by the National Soft Drink Association. It turns out that the soft drink association paid the ADA $25,000 for the soda "fact sheet," which appeared as an advertisement in the back of the *Journal of the American Dietetic Association*. The fact sheet was also posted on the soft drink association's website, under the ADA's banner, leaving the impression that the dietetic association endorses soda consumption.

The soft drink association has gone so far as to reference the ADA fact sheet in a press release defending soda from consumer advocates—without noting that industry money paid for the information. In fact, it refers to the advertisement as the ADA's "official Nutrition Fact Sheet" on soda.

When the issue began raising eyebrows, the ADA removed the fact sheet on soda from its website. The organization issued a statement in January 2004 saying that fact sheets are different from ADA positions. The dietetic association maintains it has a strict editorial policy free of industry influence. "The ADA works with industry and other sponsors to produce and fund the distribution of nutrition fact sheets," ADA guidelines say. "ADA has final editorial control of the fact sheet content."

The Center for Consumer Freedom also joined in the vigorous defense of soda. An editorial on the center's website pointed out that *The Lancet* study didn't prove soda actually *caused* the weight gain, only that kids who drank soda were more likely to put on weight. The site also referred to another study conducted by researchers at the Center for Food and Nutrition Policy (CFNP) that found no link between soda and obesity in children ages twelve to nineteen. The editorial failed to mention that this particular study was funded by an unrestricted grant from, naturally, the National Soft Drink Association.

The CFNP study, published in the obscure *International Journal of Food Sciences and Nutrition*, didn't follow children over time; instead, it compared children's weights with their soda habits. The study found that thin kids drink a lot of soda too—an interesting observation to be sure. But such observations can't come close to proving—or disproving—cause and effect. The real question is not whether skinny kids drink soda, it's what will happen to those kids two or three years down the road.

Still, the study energized a beleaguered industry. For the first time, they had at least some allegedly scientific support for their crusade to put vending machines in every conceivable location, including middle schools and high schools. An editorial in the June 2003 issue of *Beverage Industry* magazine ended with this call to arms: "The beverage industry has spent countless hours defending soft drinks and their right to be in schools. Perhaps the information [provided by the study] can help finally put that issue to rest, and also help companies work with schools to make sure kids get all of the beverage opportunities they need."

As an aside, an interesting thing happens when kids don't take full advantage of their "beverage opportunities": they stop getting fat. In 2004, British researchers published the results of an incredibly simple—and incredibly effective—effort to curb childhood obesity. A nurse visited classrooms and encouraged school kids ages seven to eleven to drink less soda or, in British parlance, "ditch the fizz." The message, delivered in just four one-hour sessions, was straightforward and uncomplicated. Kids were told that too much sugar isn't healthy and that soda could be bad for their teeth, and that's it. To drive the message home, the nurse passed around a real tooth that had been soaked in soda until the enamel started to peel away.

Not surprisingly, kids started to cut down on soda, and that one simple change made a huge difference. The obesity rate for kids who participated in the program dropped slightly over the next year. For kids who didn't hear the message, however, obesity rates climbed 7.5 percent in just one year. In this case, putting the brakes on an epidemic was as easy as "ditching the fizz."

The soda industry has certainly been under siege lately, but it has an answer—or at least a retort—to every new threat. California State Senator Deborah Ortiz's proposal for a two cent-per-can "soda tax" to fund school health programs in 2002 inspired an editorial in *Beverage Industry* that ended with this memorable line: "Hey, Deborah, stop feeding your kids Twinkies, shut off the TV, and send your kids outside to play. Maybe then you will understand what the real problem is." Thanks in part to heavy lobbying from Big Soda, the proposed tax was defeated.

Criticism about soft drinks has also riled up the sugar industry. In 2002, a law firm representing the Sugar Association sent a heavy-handed letter to Marion Nestle, the NYU nutritionist. The letter accused her of making "numerous false, misleading, disparaging and defamatory statements about sugar." One of her crimes was to suggest that soft drinks were loaded with sugar. "As commonly known by experts in the field of nutrition," the letter stated, "soft drinks have contained virtually no sugar (sucrose) in more than 20 years." In her written response, Nestle explained what *really* is commonly known by experts in the field of nutrition: fructose is a type of sugar, and soda has plenty of it.

The fast food industry is on the defensive too. McDonald's, Wendy's, Arby's, and other restaurants have started offering salads and "low-carb" options, items that help them tap into a new interest in health foods while giving them a chance to deflect criticism. It should be noted, however, that some of the new "diet-friendly" foods aren't exactly light fare. The Wendy's chicken BLT salad with croutons and honey mustard dressing has more fat and calories than a Big Mac. The Carl's Junior Low-Carb Six Dollar Burger weighs in at 490 calories and has more fat than two Snickers bars. McDonald's new Fiesta Salad with sour cream and salsa, much touted by the company, has 450 calories, or twenty more overall calories and sixty more calories from fat than a Quarter Pounder. Even if you hold the sour cream, the salad still has more calories from fat than the Quarter Pounder.

Shortly after the release of the documentary *Super Size Me* in 2004—a film in which filmmaker Morgan Spurlock gained twenty-four pounds by eating at McDonald's three times a day for a month—McDonald's started phasing out its "supersize" option. Spurlock particularly emphasized supersizing by making it a rule to always accept a supersized meal when it was offered—and in his experience, it was offered a lot. Spokespeople for the company have repeatedly said the phaseout was not a response to the movie. In other news, spokespeople have been generally quiet about plans to release a children's exercise video featuring Ronald McDonald in 2005.

In the boldest defensive strategy yet, fast food companies fought for passage of the Personal Responsibility in Food Consumption Act, better known as the "Cheeseburger Bill." This legislation would protect fast food outlets, grocery stores, and the rest of the food industry from obesity-related lawsuits. The U.S. House of Representatives passed the bill by a wide margin in March 2004. (The U.S. Senate was still considering a similar bill as this book went to press.) A few such lawsuits have actually gone to court—including a highly publicized case in which two overweight teenagers sued McDonald's—but none was successful. If the "Cheeseburger Bill" becomes law, no such lawsuits could even be filed in court, let alone be decided against the food industry.

In courts of law and the court of public opinion, the fast food industry has constantly repeated the same message: restaurants can't be held responsible for the poor choices of their customers. "That's the industry line: it's not our bad food, it's people's lack of responsibility," says Michael Lowe, a professor of psychology at Drexel University in Philadelphia and an acclaimed weight loss expert. "It's not necessarily fair to sue food companies for doing what a capitalist society encourages them to do, but they'd better show some responsibility of their own."

The defenders of fast food, soda, and sugar often say there's no *proof* that any of these foods contribute to childhood obesity, and they're right. In order to definitively prove that fast food (or soda or sugar) makes kids fat, researchers would have to recruit thousands of children for an unprecedented study. Half of the kids, chosen at random, would be told to eat fast food (or soda or sugar) regularly. The other half, the control group, would have to steer clear of the stuff for the duration of the study. ("OK, you're in the control group, so no hamburgers, french fries, or pizza for five years.") In other words, it's not going to happen.

But the theory that these foods might not contribute to obesity rests on shaky scientific footing. Take fast food, for instance. There are two ways kids can get away with those extra 187 calories that come on days when they eat fast food. They can eat less than usual on other days, or they can get more exercise. Both options seem highly unlikely, and indeed there's no evidence that kids actually make such adjustments. Likewise, there's no evidence that kids do anything to offset all of those calories in a twenty-ounce bottle of soda. They don't eat a lot less at their next meal, and they don't get a sudden urge to go walking for two and a half miles, which is what it would take to burn those extra soda calories.

Anyone who still wonders if soda and junk food really matter should take a close look at the kids at the mall food court. "What we are actually seeing is that child obesity is increasing," says Bowman, the USDA nutritionist. "That's the proof in the pudding."

A Matter of Taste

For all of the tension these days between nutrition activists and food industry boosters, there's something upon which they would all agree: a fudge brownie covered in gummy bears is just plain wrong. Only a child could even think of such a concoction without developing immediate and severe tooth pain. Kids have their own ideas about what's good to eat, and there's not much we can do to talk them out of it, Lowe says. "Children are doing what they are designed to do: eat. And if it's more palatable, eat more of it."

Very small children have a remarkable ability to regulate their calorie intake, Lowe says. When infants are given watered-down formula, they'll keep sucking until they get their full meal. And if the formula is a little extra concentrated, they'll stop a little early. "After age two or three, children largely lose that ability," he says. Some kids maintain a better sense than others about what their bodies need, but even the best instincts can be easily derailed, he says.

Many people believe children are born with a natural desire to choose healthy foods. Indeed, this theory, based on the assumption that the body knows what's good for it, was practically dogma among child psychologists in the 1960s. This concept, often called "the wisdom of the body," is so appealing that even 468 Golden Corral restaurants haven't been able to completely quash it. But once you've seen kids (and their parents) belly up to the buffet, it's hard to believe that anyone is getting really top-quality advice from their bodies. Young kids may know how many calories they need, but their instincts don't show them a healthy way to get those calories.

As Leann Birch of Pennsylvania State University recently explained in the *Annual Review of Nutrition*, there isn't necessarily much wisdom behind our food preferences and cravings. Our desires for food are a reflection of many complex factors, including the hardwiring in our brains and the food we eat early in our childhood. Kids are born with a fondness for sweet tastes, an aversion for bitter and sour tastes, and a reluctance to try anything new. (Children also have an inborn liking for salt, but this doesn't show up until they are several months old.) The preference for sweets runs especially deep, and it starts early. Newborns drinking a sugar solution look "happier" than newborns drinking plain water. A sip of sugar water can even soothe a baby boy during circumcision!

Beyond these basic tendencies, children are largely a blank slate, according to Birch. As they grow, their tastes are shaped by the foods they are offered. When kids are given healthy foods at a young age, they can learn to like them. But broccoli and grilled cheese sandwiches don't compete on an even playing field. If given a chance, kids very quickly learn to prefer foods packed with calories—probably a reflection of our evolutionary past when food was scarce and children needed every calorie they could get in order to grow.

The tendency to horde calories served children well in the past, but it's working against them today, Lowe says. "Why are obesity rates going up? You don't have to look beyond the environment: food is omnipresent, inexpensive, and very palatable."

The Age of Ads

The food industry is making a killing from kids' hardwiring, but that's not enough—the industry is also trying to rewrite their programming. According to a 2004 report from the American Psychological Association, advertisers spend $12 billion a year on ads aimed at kids, $3 billion of which comes from fast food companies. For comparison, the government spends $3.6 million each year on its Five-A-Day campaign that promotes fruits and vegetables. The average kid watches 40,000 commercials in a year, twice as many as in the late 1970s. The *Handbook of Children and the Media,* published in 2001, reported that thirty-two percent of ads aimed at kids are for candy, thirty-one percent are for cereal (often sugarcoated), and nine percent are for fast food. All told, kids see about eleven food commercials per hour of television, and not a single one of those ads heralds the hottest new apples or carrot sticks.

Adults may scoff at ads—especially the ridiculously cartoonish fare that dominates children's television—but kids have a different reaction. In fact, young kids may not realize that they're watching an ad at all. A task force from the American Psychological Association reported in 2004 that children under eight generally don't understand the purpose of advertisements. They think ads are truthful and straightforward, as if the good folks at Nabisco or Hershey's just wanted to make sure everyone knew about the latest cookie or candy bar.

"Because younger children do not understand persuasive intent in advertising, they are easy targets for commercial persuasion," said Brian Wilcox, PhD, chair of the APA task force, in a release that accompanied the report. "Advertising of unhealthy food products to young children contributes to poor nutritional habits that may last a lifetime and may be a variable in the current epidemic of obesity among kids."

Kids certainly take those thirty-second "informational" messages to heart. As any parent knows, ads on TV quickly turn into demands at the grocery store. Even toddlers will plead for foods that they see advertised on TV or that feature a favorite television character. (Chris Woolston, one of the co-authors of this book, can confirm that his three-year-old goes berserk whenever he sees a box of Blue's Clues macaroni and cheese. The store thoughtfully puts the boxes right at his eye level so he doesn't have to strain his neck.) Television can also warp a kid's sense of nutrition. When fourth and fifth graders were asked to pick the healthiest food from two similar options (such as corn flakes and frosted flakes), the kids who watched the most TV were most likely to make the wrong choice.

The reach of child-focused marketing goes far beyond television, Ebbeling says. Fast food companies have been especially innovative when it comes to attracting children's attention. "When I was a kid, there weren't even big playgrounds at McDonald's," she says. "Now you can go to a Toys"R"Us and buy clothes to help you pretend that you work at McDonald's. Play-Doh has toys that make fries and burgers. You can even dress up Barbie as a fast food employee." It's still not exactly clear how Barbie can afford her dream house.

The Internet is opening up whole new frontiers in the battle for brand loyalty. Kids who visit nabiscoworld.com—a site promoted on food packages, television ads, and web portals such as Yahoo—can play dozens of edifying games that all feature Nabisco products. For instance, they can drive an armored car full of Barz (a chocolatey combination of a cookie and a candy bar) around city streets on a mission to deliver their precious cargo to as many stores as possible before time runs out. After a few rounds in an armored car, a kid could be forgiven for thinking that Barz are about as valuable as gold bullion.

The Dieting Dilemma

Asking children to eat well in this world is like asking them to go out in the rain without getting wet. No amount of training or cajoling can completely protect them from the downpour. Kids can respond to simple, direct messages—as they did in the "Ditch the Fizz" study—but they can't take on the burden of healthy eating entirely by themselves. You can give a fifteen-year-old a mini-PhD in nutrition. You can teach him

Clay's story: "I wanted to be less embarrassed"

Moving to a new school is never easy. But when you're seventeen years old and more than one hundred pounds overweight, the prospect of walking down a hall surrounded by new faces takes on a new level of terror. Clay Jones of Enterprise, Mississippi, had known for years that he needed to lose weight. He tried Weight Watchers—and failed. He saw a nutritionist who taught him how to count calories—and got nowhere. His weight passed 280 pounds, and 300 didn't seem far away. But when he transferred from his private school to a public school in his junior year of high school, his desire for a thinner body became a mission. "I wanted to be more social and less embarrassed," he says.

Clay is twenty-four now, but he is still transformed by a choice he made in high school. He decided to take one more stab at changing his diet, and he was going to do it his way. It was a little extreme, and it certainly wasn't what the experts recommend, but it worked. "I did the opposite of Atkins," he says. "I started to count fat grams." At first, he limited himself to just ten or fewer grams of fat each day, which is equivalent to about half of a McDonald's Quarter Pounder. "I can tell you how many fat grams are in any food," he says. He ended up eating a lot of corn flakes, as in four or five bowls at a time, with skim milk, of course. He dropped thirty pounds within three months, and by the time he graduated from high school, he had made it all the way down to 160 pounds.

Clay couldn't stick to his super-restrictive diet forever. He slowly started to add fat back into his meals, and sure enough he started to gain weight. But he hadn't forgotten what he'd learned. He still knew the hidden places where fat can lurk, and he knew what he could eat without much risk. He went up to about 180 pounds—a hundred pounds shy of his heaviest weight—and he has held steady ever since.

Looking back, he has a hard time seeing how he ever got so large in the first place. He doesn't remember eating more than anyone else around him, and he wasn't exactly stuck in front of the television all day. He played summer baseball and soccer in grade school, but his body kept getting bigger and slower. "I think I had a predisposition for a slow metabolism," he says. By the time he was in fourth grade, he had spent a lot of time "riding the bench" on the baseball team. "I told my coach that I wasn't going to run the bases in practice because I wasn't getting in the games." When this comment got back to his mother, she personally chased him around the bases and gave him the distinct impression that it was in his best interests not to get caught.

Clay doesn't talk much about his "fat days," but he still carries a reminder. "I have a lot of excess skin," he says. His past was in full view when he recently went swimming with a new girlfriend. For the first time, she got a good look at the stretch marks on his arms and stomach, and she didn't seem to mind. Some people might look at that extra skin as a physical flaw. But it could just as easily be seen as a badge of honor.

to read food labels. Go ahead and give her a portable food scale and a set of calipers. Chances are you'll end up with a teenager who happens to be unusually knowledgeable about the hamburgers and pizza he consumes constantly.

Nutrition education is important, to be sure. No child should be let loose in the world until she knows the difference between calcium and cholesterol, and no one should be under the illusion that a Big Mac and fries makes for a healthy meal. But education can only go so far. For the most part, kids can't simply "learn" to eat healthier food.

Even those who try putting themselves on a diet rarely manage to cut calories in the long run. Researchers at Harvard Medical School recently tracked the dieting habits of more than 16,000 boys and girls ages nine to fourteen. Nearly thirty percent of the girls and sixteen percent of the boys went on a diet over the course of three years. As reported in *Pediatrics* in 2003, girls who frequently dieted were twelve times more likely than nondieting girls to go on food binges. Among boys, frequent dieters were more than seven times more likely to binge. Overall, the kids who put themselves on diets ended up gaining more weight than the kids who didn't diet.

Of course, not all "diets" are created equal. At one end of the spectrum, some kids go on starvation diets and put themselves at risk for eating disorders. At the other end, some young people make sensible, sustainable changes at mealtime that really improve their health. As Lowe recently wrote in the journal *Obesity Research*, "Asking whether dieting is desirable is akin to asking if taking drugs is desirable (when 'drug-taking' can range from prescribed medications for an illness, to drinking socially, to mainlining heroin)."

These days, you can't talk about dieting without talking about low carbs. Many older kids and teenagers have hopped on the Atkins bandwagon, but is it a good idea? To date, very little research has been conducted on the long-term safety and effectiveness of low-carbohydrate diets for weight loss in children and adolescents. Doctors sometimes prescribe ultra-low-carb diets to kids to control seizures, an approach that does have some reported long-term side effects, including kidney stones, calcium loss in bones, high cholesterol, and heart trouble. Low-carb diets often lead to rapid weight loss, but most people eventually return to a normal diet and regain all or most of the weight. The long-term results are just too murky to recommend the Atkins diet or any other low-carb approach for the younger set.

Older kids who want to try a meal plan that can help them slim down without any risks may want to examine another approach. Ebbeling and other researchers at Harvard recently conducted a small study of "reduced-glycemic-load" diets on sixteen overweight adolescents ages thirteen to twenty-one. Like the Atkins diet, the reduced-glycemic-load diet cuts down on the refined starches and sugars that quickly send blood sugar soaring. (The term "glycemic" refers to blood sugar levels.) But unlike Atkins, the white rice, white bread, and muffins aren't replaced with bacon, eggs, and steaks. Instead, the subjects got their carbohydrates from healthier sources such as non-starchy vegetables, fruits, and whole grains.

"There are a couple of different ways to reduce glycemic load," Ebbeling says. "You can reduce the total carbohydrate or reduce the type." Carbs made up forty-five percent of the calories in the study diet, which is far more than Atkins would allow but still less than the typical American diet.

Researchers put the kids on the diet for six months and then checked in with them six months later. As reported in 2003 in the *Archives of Pediatric and Adolescent Medicine*, the eating plan was an unqualified success. A year after the study started, kids on the reduced-glycemic load plan had lost more weight and more body fat than other kids who were put on a traditional low-fat diet. The reduced-glycemic load plan is "healthy and risk free," Ebbeling says.

Shootout at the Golden Corral

In the big picture, any approach to childhood obesity that hinges on a kid's ability to count calories and sort "good" foods from "bad" is almost certain to fail, Lowe says. The never-ending mantra of the fast food industry won't do any good either, Ebbeling says. "Fast food companies say it's a matter of personal choice," she says. "But when you're dealing with a child or an adolescent, relying on personal choice is going to be difficult."

Instead of nagging our kids to make better choices, we should ensure that they have only good things from which to choose. It's time to take a hard look at all of the forces that encourage them to choose badly and to overeat. "We may not be able to change people, but we certainly know how to change the environment," Lowe says. "It's far easier to maintain changes in the environment than changes in people. We know what needs to be done, but powerful forces are against it." If we're really serious, Lowe says, we should support establishing subsidies to lower the price of healthy foods, getting rid of soda and candy from schools, encouraging food manufacturers and restaurants to provide more low-calorie options, and requiring restaurants to clearly list the caloric content of every item on the menu.

In other words, it will take a revolution. Parents, public health activists, and legislators are already gathering the troops, but today's kids can't wait for the future victories. For them, the food environment has to change now, and the change has to start at home. Unfortunately, in this age of fad diets and convenience food, many parents have lost touch with what kids really need at different stages of life. Good nutrition isn't really all that complicated, and it certainly isn't out of reach. Kids who eat a wide variety of healthy foods each day tend to stay trim and healthy, and they'll develop tastes and habits that will serve them well throughout their lives.

Supersize portions, the drift away from the family meal, the overabundance of calorie-packed foods, the ubiquitous ads for the latest sugary and fatty snacks—with so many forces working against kids, we'll all have to do more than count calories. The American diet may be getting more childish, but adults can change things for the better. At last check, we were still the ones in charge.

4

Schools: From Sellouts to Sanctuaries

WHEN NEWLY HIRED Aptos Middle School principal Linal Ishibashi walked into the school cafeteria for the first time in fall 2002, she wasn't expecting four-star cuisine. But she didn't anticipate a full-scale nutritional nightmare either. At least half of the kids at the San Francisco school were making a lunch of a bottle of soda and an extra-large bag of chips from the school lunchroom. The rest of the students were enjoying chicken wings, hot links, mega-colossal cheeseburgers, and pizzas that could feed a whole family.

Ishibashi looked at this mess, all purchased at her school cafeteria, and imagined her own daughters eating it. Although her children were "very well-trained about food," she says, she wondered what would happen if they were exposed to the same environment. Would they be able to make the right choices? And if not, why would she expect that anyone else could? She quickly placed a call to the nutrition department to ask if she could please purge her school of soda and chips. No, she could not. Those items, she was told, made too much money for the district.

In high schools not far from Aptos, students were loading their trays with Taco Bell tacos, Subway sandwiches, and Domino's pizzas, all available in the school cafeteria. They were heaping on fries and chips, pastries and cookies. After lunch period, many of them washed it all down with a twenty-ounce soda from a vending machine just outside the lunchroom.

The last thing American kids need, of course, is more junk food. One in three kids in the United States over the age of four eats fast food every day, and according to a study published in the journal *Pediatrics* in January 2004, kids who eat fast food daily get enough calories to pack on six extra pounds each year. Yet we feed them this fare in school, where the government has vowed to take care of them. According to a national survey done in 2000 by the CDC, more than twenty percent of schools sold McDonald's hamburgers, Pizza Hut pizzas, or other brand-name fast foods. A separate survey of California high schools, also done in 2000, found that ninety-five percent sold fast foods à la carte—that is, separate from the main school lunch and any government scrutiny—including many foods from big-name franchises.

School budget woes have opened the door to a veritable feeding frenzy on school property. Anything that local school boards, principals, and school superintendents allow—sometimes over the objection of food service directors—can be sold in school. That, of course, includes soda.

For years, soda companies have been signing exclusive, largely secret contracts with school districts to install vending machines on school property in exchange for cash. In 1999, when the practice really began to escalate, a Coca-Cola spokesperson described this arrangement as a "win-win-win" for schools, for thirsty kids, and for Coca-Cola. Soda companies win loyal young customers; financially strapped schools make money. Many school districts reasoned that kids would bring soda from home anyway. With vending machines no more than a few steps outside their classrooms, they could have their soda and give money back to their schools. Few school districts stopped to debate whether increasing the students' exposure to soda could be harming their health.

As far as the nation's pediatricians are concerned, the verdict is already in. In January 2004, the American Academy of Pediatrics issued a policy statement warning children's doctors that obesity can be associated with a high intake of sweetened drinks. The group noted that sweetened drinks are the primary source of added sugar in kids' diets and that soda often takes the place of more nutritious, calcium-rich drinks such as milk. The physicians' organization called on schools to consider restricting sales of soda. It also encouraged pediatricians to lobby against making soda available in school, something that members of the association have taken to heart. In Ohio, pediatricians declared that soda had no place in the schools. "The presence of a vending machine in a hall outside the classroom," they noted, "is a far more powerful statement than whatever is said within the classroom."

According to the 2000 CDC survey, more than seventy-six percent of public schools in the United States sell soft drinks from vending machines. Kids can buy soda in nearly ninety-four percent of high schools and fifty-eight percent of elementary schools, according to a Robert Wood Johnson Foundation (RWJF) report. In exchange for the exclusive right to sell their products, soft drink giants such as Coca-Cola and Pepsi return a portion of the profits to underfunded schools. These funds can pay for such essentials as textbooks as well as sports and music programs. Schools get band uniforms and Big Soda gets brand loyalty.

Soda companies are even shaping policy at one of the nation's most venerable school organizations. In 2003, the Parent Teacher Association accepted sponsorship by Coca-Cola and gave Coke Enterprises Senior Vice President of Public Affairs John H. Downs Jr. a seat on the PTA's national board with full voting rights. The PTA's defense is that Coca-Cola is endorsing them, not the other way around.

This sort of arrangement between marketers and educators has escalated with jaw-dropping ferocity. In September 2004, Atkins Nutritionals Inc., purveyor of the popular high-protein, high-fat diet, announced it would team up with the National Education Association, the nation's largest teachers union. Atkins said it plans to help pay for NEA's school health website. A spokesman for the NEA told USA Today that Atkins was not intent on steering children toward high-protein diets, but critics noted that the deal will bring the Atkins brand name into schools and be worth millions of dollars in advertising. "Obviously this is nothing new, but it's much more profligate and pervasive than it's ever been," Arnold Fege of the Public Education Network told the newspaper. Fege, whose group advocates for low-income children, went on: "I think where it crosses the line is where you have a *quid pro quo*, where the schools become part of the marketing department of the corporation."

School is supposed to be a place where we teach children about critical thinking and personal responsibility. Yet in many schools, kids hear one message in class and see another in the lunch line. We teach them about nutrition and the importance of making good choices, yet we surround them with bad choices. We teach them the value of thinking for themselves, yet in their own institutions of learning, we bombard them with posters, banners, and scoreboard advertising for Coke, Pepsi, and a range of fast foods—all of which, consumed in excess, raise the risk of obesity and all its attendant ills. "The question is, why are we harming these children?" asks Alex Molnar, professor and director of the Education Policy Studies Laboratory at Arizona State University. "Are we saying these children have to be sick to be educated?"

Welcome to Main Street School, Sponsored by McDonald's

How did schools become "partners" with Coca-Cola, Pepsi, McDonald's, and Taco Bell in the first place? It's the same sort of symbiosis that allows soft drink and food companies to sponsor the American Dietetic

Association or become members of the School Nutrition Association, which represents school cafeteria managers. It's about the money. The ADA, the SNA, and other professional and advocacy organizations say they need corporate sponsors to help fund their programs and that their partners don't influence their approach to nutrition. Soda partnerships are worth enormous amounts of money, in some cases millions of dollars per district, although they often don't amount to much per school. Schools desperately need more funding—something the soft drink industry has long understood.

The groundwork for the school feeding frenzy was laid more than thirty years ago. Standards for the nutrition content and portion size of school meals have always been the province of the USDA. Wary of soda and vending machine foods encroaching on schools, Congress amended the 1966 Child Nutrition Act in 1970, allowing the USDA to regulate foods that competed with the school lunch program. But schools making money selling snack foods objected, and in 1972, Congress rescinded USDA's authority. "Profit had triumphed over nutrition," a U.S. District Court judge would later write.

Following two years of hearings and 15,000 pages submitted in evidence, Congress reversed itself. Soda, water ices, chewing gum, and certain candies could be eliminated from sale at schools. But the National Soft Drink Association roared back with a lawsuit in 1980. It lost the suit in 1981 but won on appeal two years later. The U.S. District Court of Appeal agreed with the soda companies that the Child Nutrition Act didn't give federal regulators authority to regulate the time and place of soda sales, although it did agree that "a vending machine, no matter where located, can operate as a magnet for any child who inclines toward the non-nutritious." In other words, the law was on Coke and Pepsi's side, and schoolchildren would have to fend for themselves. It meant that—along with soda—candy bars, cookies, doughnuts, and potato chips are still defined as reasonable foods for sale in schools.

Carol Tucker Foreman, who was a top official in the USDA until shortly before the lawsuit rulings, says that throughout the 1980s the Republican-controlled Congress had little desire to regulate competitive foods in schools. "For twelve years, under the Reagan and Bush administrations, the Department of Agriculture was fully in the hands of the food industry," says Foreman, who is now director of food policy for the Consumer Federation of America. The Clinton administration missed the opportunity during the first four years. But by Clinton's second term, Congress was again dominated by the Republican party, and nothing was done. "It was one of the most deplorable capitulations on public policy that I can think of, with all of the knowledge we have, not to drive the junk food out of schools," says Foreman.

But in some ways it was just politics as usual. The food and agribusiness industries have enormously powerful lobbies and bountiful cash. In 2002 agribusiness, which includes such companies as Phillip Morris, RJR Nabisco, PepsiCo, and U.S. Sugar Corp., made $54.3 million in campaign contributions; in 2000 the food and agriculture industries together spent $77.5 million lobbying in Washington. Not surprisingly, for three years' running since 2000, bills to redefine "foods of minimal nutritional value" sold in schools have failed to get out of committee.

"Nobody stands up and says, 'We want kids to drink more soda in school, and we want kids to have as much candy as possible,'" says Margo Wootan of the Center for Science in the Public Interest. "They hide behind other arguments like 'local control.' All this talk about local control is hogwash. It's a smokescreen to keep junk food in schools."

What every parent wants—and what few get these days—are schools that are inviting, cheerful places, with sunlight streaming through the windows, where no teacher has to scrounge for pencils, paper, or textbooks, and where music classes and physical education are not considered "extras." To be sure, such classrooms still exist in public schools, but for many, sunlight streams in only because the window shades don't work anymore, and there's no money to fix them. Lockers that stay attached to the wall or toilets that flush—let alone band uniforms—are beyond the reach of many schools.

The lack of funding is also apparent in the dismal places in which many children are expected to eat lunch. At Samuel F. Gompers Continuation High School in Richmond, California, for example, teens not eligible for a free lunch must queue up to a window and select from a virtual hypertension menu of regular soda (no diet sodas), chips, or salt- and fat-laden noodles in Styrofoam cups. Because there's no school cafeteria, they have to eat this fare while standing in the hallway.

Parents across the country have protested the abysmal state of our public schools. But every year, as funding tightens, school administrators struggle to meet their classrooms' most basic needs. So, often over the pleas of school dietitians, superintendents and school boards tend to say "yes" when Coke and Pepsi come knocking.

Food-service directors are also under enormous pressure. To keep from going over budget, many school food-service departments are told to buy the cheapest bulk foods, leaving out more expensive fresh produce. With the hope of keeping student customers, they'll also sell competitive foods, also known as à la carte items, from fast-food chains. Although the federal government controls the nutrient and fat content in the regular school lunch, à la carte items are beholden to no one, except possibly Ronald McDonald.

No matter what their motives for selling fast food and soda, schools are violating a sacred social contract, says ASU's Molnar. "Schools are set up by our society as protected spaces. We use the police power of the state to require people to be there. There are only two other institutions in our society in which that's true—prisons and mental hospitals," Molnar told members of the SNA in January 2004. "It's a war zone in classrooms right now with regard to nutrition. Anybody and everybody that can promote a product or a service (related to food) is in the classrooms. Not just in the lunchrooms, but in the classrooms, in the hallways, on the buses."

Fast food, candy, and Big Soda are also starring on TV screens in schoolrooms across the United States. Channel One, which broadcasts overseas and to 12,000 schools in the United States, boasts an audience of eight million American sixth to twelfth graders. The television channel, owned by PRIMEDIA and begun by New York public relations firm Whittle Communications, lends a television to each classroom in subscribing schools, along with VCRs and satellite transmission (connecting only to Channel One). In exchange, schools must tune in to Channel One's news programs for kids on most days in most classrooms for at least twelve minutes a day. That may not sound like much. But when you add up the time that kids are captive in class with the TV blaring, it comes to one lost week of school per year—or seven weeks over the course of seven school years.

Consumer advocate Jim Metrock says that Channel One's advertising for everything from M&M's to Snickers bars to McDonald's to Lay's potato chips is relentless. Young "reporters" hawk products (such as newly released CDs) on the air, leading some parents to wonder what kind of "news" their kids are getting.

"We have public schools, government institutions, all luring our children to eat more Twinkies and drink Pepsi," says Metrock, whose organization, Obligation, located in Birmingham, Alabama, is dedicated to fighting school commercialism. Metrock, who spent his life as a businessperson, is by no means an anticapitalist radical. He used to own a steel company and made a fair profit from it. But commercialism aimed at schoolchildren always sickened him. After he sold his business, he decided he wanted to do something that would remind other businesspeople not to put money before the welfare of children.

Some school districts already share his views, and the states of California and New York have refused to let Channel One into their schools. Others, such as the 20,000-student district in Shelby County, Alabama, have soured on the company and cancelled their contracts. Those who continue with Channel One say it's a small price to pay for equipment they would never be able to afford. But Metrock believes in the old adage that no lunch is free.

"People have got to get more involved with their public schools, and I don't just mean parents; seventy percent of taxpayers at any one time don't have kids in public schools," says Metrock in his mild Southern drawl. "To guarantee these people an audience so that they profit off that—we ought to take them to the woodshed."

Canning Soda

Some parents, teachers, public policy advocates, and school boards are listening to Metrock's advice—they're taking the food and beverage industry and their exclusive contracts to the proverbial woodshed. The Los Angeles Unified School District, the second largest school district in the nation, became a leader in this movement when the school board unanimously banned soft drinks and sugar-laden fruit drinks beginning in 2004 for all grades at all campuses. "It's a health issue," said Rosemary Lee, a teacher at Logan Street Elementary School in L.A., who spoke in favor of the soda ban. "I have kindergartners who are ninety pounds at four or five years old. I have kids who come to school with a soda for breakfast."

The Los Angeles decision paved the way for the California legislature to make some history of its own—banning soft drinks from elementary and middle schools (although not in high schools) also as of 2004. The Los Angeles decision was a major feat that began with one teacher. Jacqueline Domac, a health teacher at Venice High School, was dismayed about students' easy access to soda at her school. When she investigated further, she said she was horrified to learn that her school's exclusive multimillion-dollar contract with Coke had to be kept secret and confidential. Domac and her students found, not surprisingly, that ninety percent of the beverages sold in Coke's vending machines at their school were considered unhealthy by state nutritional standards.

Getting rid of soda would have been a slam dunk were it not for worries about what a soda ban would do to the financially troubled school district. "I think that everyone is interested in providing healthy foods for kids, but some of my colleagues were very concerned about what the schools would do if they lost funding," school board member and soda-ban cosponsor Marlene Canter said after the vote. "I didn't think we should mix up the economics of this because this is a health issue, not an economics issue. Otherwise we'd be saying it's okay to sell unhealthy things as long as they made money for our schools." It took a passionate three-hour debate, but in the end, the Los Angeles board voted unanimously to give soda the boot.

Interestingly, Wootan's organization found that although it sounds like a lot of money, what districts get from soda sales doesn't come to that much in the end. "The amount of money that schools are earning out of vending machines is very modest," Wootan says. "The most lucrative soda contract we found raised $30 a student per year, and it's less than 0.1 percent of the school budget. It sounds like five million is a lot, but five million over five years often comes to about $50,000 a year per school."

In fact, when she closely compared the numbers, Texas Agriculture Commissioner Susan Combs made an even more revealing discovery. Although Texas schools make $54 million in revenue from vending contracts, the state's schools *lose* $60 million in meal sales every year in competition with vending sales. If every state looked at the numbers that closely, vending contracts might not look so lucrative after all.

Vending machine food didn't seem to Combs to be a good deal for kids—nutritionally or financially. Over the objections of the Texas Association of School Boards and the Texas Association of School Administrators, she implemented what may be the most progressive statewide school food policy in the nation. Beginning August 1, 2004, Combs banned sales of candy, soda, and other sweets in elementary schools until after school. The policy banishes competitive foods from middle and high schools during lunchtime. No soda sold in high school can be bigger than twelve ounces, with the goal for the 2005–2006 school year that no more than thirty percent of the space in vending machines be occupied by soda.

Other states have made significant changes as well. The state of New York now prohibits the sale of candy and soda from school vending until the last lunch period. (The New York City school system—the nation's largest, with about one million children—banished soda from its schools only to sign an $8 million-a-year contract making Snapple the official drink of New York schools. Although the company will be selling 100 percent juice and bottled water alongside its fruit drinks, the sugar-sweetened drinks are just as bad for kids as soda, according to Marion Nestle. Nestle told the *New York Observer* she was "appalled" by the controversial decision.) Arkansas prohibits vending machines in elementary schools, and middle and high school students don't have access to the machines until after lunch. Legislators in at least nineteen other states are trying to restrict kids' access to soda in schools. Local school boards as well have started to chip away at Big Soda's share of the school market.

Even in places where soda is still welcomed by the school district, conscientious school nutrition directors have been able to set limits. Rick Mariam, director of Food and Nutrition Services in the Pattonville School District near St. Louis, Missouri, couldn't stop his district from signing a relatively small $150,000 contract with Pepsi, but he did manage to limit students' access to vending machines to before and after school. Still, Mariam's headaches continue: elementary school teachers have soda machines in their lounges and often reward kids with a trip to the machine for doing good work. "Anything we do in the schools is putting a stamp of approval on it, so when we sell sodas, we're saying it's a healthy product," says Mariam.

A twenty-five-year veteran of school food service, Mariam says the soda marketers come into his office with flip charts showing they're not making much profit, especially on non-soda beverages, but they tell him that the brand loyalty they're building in children makes up for it. "It touches a nerve in me," says Mariam. "They should be held responsible. They shouldn't be allowed to market to students. They're serving a health-hazard product, and they should be responsible for the (health) costs incurred."

"Partners Don't Let Partners Do Bad Things to Kids"

In its Nixon-era heyday, the Hyatt Hotel in Newport Beach, California, hosted so many dignitaries and Nixon cronies it was known as the western White House. On a breezy day in January 2004, several hundred members of the School Nutrition Association gathered at a glittering annual awards ceremony in the hotel's Plaza Ballroom. The yearly event honors the achievements of its members—public employees who manage food programs at public schools. The award presenters and hosts included Campbell's, Tyson Foods, and Basic American Foods. In accepting their awards, each honoree made sure to thank the food companies personally. The title of the conference itself—Building Business Partnerships—said much about its purpose.

The next day, the group held a panel discussion on the school nutrition environment, which included Cheryl Sturgeon, a food service director from Kentucky; California State Assemblywoman Lynn Daucher (R–Brea); Barry Levenson, a former assistant attorney general who specializes in food issues and law (also a curator of a mustard museum); Joan Miller from San Antonio's health collaborative; and nutritionist Barbara J. Ivens, a pediatric nutrition specialist employed by PepsiCo Beverages and Foods. All were discussing how to improve the nutrition environment for kids. Rather than suggest removing soda from the schools, Ivens offered that overweight and obesity are complex problems not solved by any one measure. When Rick Miriam stood up and asked whether the organization might bring a class-action suit against the soda companies for selling a harmful product in schools, there followed an uncomfortable silence. What could the panelists say? The room was full of SNA members representing the food and beverage industry.

SNA President Donna Wittrock says that industry members have belonged to the organization since its inception in 1946 and make up only approximately one percent of the group's 53,303 members. Of the group's $8.5 million yearly budget, the food industry contributes about $425,000 in dues and sponsorships. Wittrock says that without industry's financial support, the organization might have to cut back on conferences and other member benefits. Industry members play an essential role advocating for better nutrition in schools, she says, and the close relationship has led to the development of healthier products, including the elimination of trans fat in some Frito Lay products. (Presumably that includes the fat reduction in the Flaming Hot Cheetos they're serving in San Antonio.) The organization "is not beholden to the food and beverage industry," Wittrock says. It has "found it better to have a dialogue and a relationship with those with differing opinions and agendas in the hopes that we will serve to educate them."

Clearly the moral dilemma of marketing to kids was on the conference organizers' minds. In a panel exploring the topic, the black-haired, boyish-looking Cliff Medney, director of concept development at East/West Creative in New York City, was pitted against the gray-bearded, professorial Alex Molnar from Arizona State. While Medney bounded back and forth across the stage, pushing buttons on his PowerPoint presentation, gesturing at his flip chart, and introducing a few tortured acronyms such as INFLUENCE—Inspire Nutritional Foods by Leveraging Understanding and Enthusiasm for kids' Natural Curiosity to Explore—Molnar stood stock still behind the podium, bereft of graphics.

Medney told the group that they were in a unique position to market good nutritional messages to kids in school, their natural habitat, to counterbalance the bad ones they are getting. "If they're informed and intelligent and smart, then kids will make the right choice. You know what's on their plates in 2004?" Medney said. "Opportunity." Judging from the response to the panel, the food service people thought Medney was taking a real-world view of what they have to deal with.

Molnar, a hard-liner in a world where the boundaries between government and commerce are constantly shifting, had a much more difficult time winning hearts and minds. He acknowledged that school people have to do business with vendors, but he warned them not to be wooed by marketers. He mused over whether, like an offer made to one of the Marx brothers in the film *A Night at the Opera*, one's assassin could also be one's bodyguard.

"I like partnerships. Nobody doesn't like a good partner," Molnar said. "But I would say to you that partners don't let partners do bad things to kids. . . . The people who will try to sell you their products and services are going to try to sell you the products and services that are most profitable to them. And they hire marketers to create demand. We now see marketing in the guise of curriculum materials, we see marketing on Web browsers, we see marketing in the form of exclusive agreements with soft drink bottlers, and we see an explosion of childhood obesity."

Molnar's opinions naturally placed him at odds with the National Soft Drink Association, which has always dismissed any connection between soda and obesity. In its policy statement on efforts to restrict or ban soda sales, the association contends that soda is part of a balanced diet that aids hydration and that soft drinks have been scapegoated in the war against obesity. In fact, the association argues, lack of exercise is a much bigger factor than soda. To back up its claim, soda giants Coke and Pepsi are pouring millions of dollars into youth athletic programs. The soft drink association notes that sixty-six percent of schools in partnerships with soda companies put their soda

profits into physical education and sports equipment—a perfect example of Molnar's assassin/bodyguard metaphor.

"I'm very glad that PepsiCo is sponsoring exercise," he said. "But you tell me. You do the math. How far do you have to run to run off two twenty-ounce Big Gulps filled with Pepsi-Cola?"

Running on Empty

School kids probably won't be able to answer that question from personal experience: fewer and fewer are running in school these days. Neither corporate sponsorship of school athletic programs nor the obesity epidemic has prevented drastic cuts in elementary and high school physical education programs across the country.

Children need at least an hour of moderately vigorous exercise a day, according to the National Association for Sport and Physical Education (NASPE), but half of all U.S. children get less than thirty minutes of exercise daily. Once a child's most reliable source of exercise, school PE classes are in danger of becoming a thing of the past.

Although the PE classes of the last few decades were not always ideal—unattractive gym suits and teasing of unathletic students abounded—they at least provided students with some daily exercise. But according to the CDC, the number of high school students taking daily physical education classes declined more than thirty percent between 1991 and 2003. Even though the CDC recommends requiring daily PE for kindergartners through twelfth graders, federal law does not mandate it. Currently, only one state—Illinois—requires daily PE classes for all school kids. Even recess—that sacred time for running, shouting, and four-square games—is becoming a luxury as elementary schools try to squeeze more academic instruction into each day.

This leaves soda companies with a ready argument: at the very moment when antiobesity advocates are flogging Big Soda, schools are shutting down PE classes. How could Coke and Pepsi, those champions of exercise, possibly be the bad guys?

"First of all, let me say, 'People, get your facts straight!'" Tom Bachmann of the *Beverage Industry* journal wrote in a scathing editorial. "The main cause of obesity in kids, and adults for that matter, is lack of exercise. Many school districts have dropped mandatory physical education from their curriculums. . . . Is this the fault of the beverage industry? No."

What Bachmann would like to see is more bottlers following the lead of John Alm, president and chief executive officer of Coca-Cola Enterprises, Coke's largest bottler, when he heard the stunning news that the Los Angeles school district had voted to get rid of soft drinks. "Enough is enough: What is the plan?" he asked his public affairs director John Downs, who landed on the national PTA board two months later. Coca-Cola Enterprises quickly decided on a strategy that emphasized promoting physical activity for kids rather than cutting out soft drinks; it also made it a policy to offer a choice of bottled water and juices along with soda and eliminated large upfront payments for long-term contracts. "I ask that each employee become an advocate," Alm said in a memo. After all, he later told the *Atlanta Constitution*, "The school system is where you build brand loyalty."

Farewell to Koolies

Although soda is ubiquitous in secondary schools, it didn't get its start there. Big Soda went to college first. In 1992, Pennsylvania State University signed a $14 million, ten-year contract with PepsiCo, giving the beverage company exclusive vending and advertising rights on each of the college's twenty-one campuses. The deal gave Pepsi an almost captive consumer base of 70,000 students. For its part, Penn State, renowned for its football team, didn't have to go begging for athletic sponsorships. That first contract, which was the most encompassing school deal with the beverage industry at the time, was the precursor to the trend we're seeing today. The university had no trouble justifying the arrangement after the legislature slashed the school's budget by 3.5 percent.

Other colleges quickly followed: University of Cincinnati, University of Oregon, Rutgers University, University of Minnesota. By the time University of Maryland signed its fifteen-year deal with Coca-Cola in 1997, one hundred college campuses had penned deals with either Coke or Pepsi. The momentum to sign up public school districts picked up two years after Penn State's agreement. Pennsylvania school districts began signing on. Then schools in Texas. Then Madison, Wisconsin, and others. But it wasn't until 1998 that exclusive deals between school districts and Big Soda really took off, according to the Center for Commercial-Free Public Education.

Not all contracts are contingent on how much soda gets sold in school, but some districts are paid by commission, pressuring schools to hawk even more soda to kids. In 1998 John Bushey, the executive director of school leadership for Colorado Springs School District 11, wrote a letter to the district pointing out that it was lagging far behind its goal to sell 70,000 cases of Coke products. Without sales in that stratosphere, it wouldn't get the full benefits from Coke. Bushey, a public school employee, was so engaged in his role promoting soft drinks that he called himself the "Coke Dude." He suggested there be almost no limits on when during the school day students could buy soda and promoted locating vending machines in areas where students would have easy access. Bushey said if necessary he could provide more electrical outlets to keep even more vending machines humming.

These days, the small but growing anti-soda backlash may likely keep school administrators from demonstrating that much enthusiasm. "These vending machines are cash cows. . . . We've had to become real scrappers in public education" to provide for students, Charles Maranzano, assistant superintendent of schools in Dinwiddie County, Virginia (near Richmond), told the Associated Press. "We've become dependent on this revenue." But he didn't sound entirely comfortable with the arrangement. "We, on one hand, want to promote good physical health . . . and yet we infuse [students] with megadoses of sugar. We probably shouldn't be doing that."

Although fast food and soda are everywhere in schools, parents and teachers overwhelmingly believe it's wrong. The Robert Wood Johnson study on school food and exercise trends found that a whopping ninety-two percent of teachers and ninety-one percent of parents said they favor converting the selections in school vending machines to healthy foods and drinks. Politicians tend to notice such majorities, so it's no surprise that state legislatures have been targeting the most egregious affronts to good nutrition. By 2004 twenty states had already restricted students' access to junk food in vending machines.

Across the nation, the switch to healthier school food has had small grassroots beginnings. In the town of Opelika, Alabama, there hasn't been a vending machine on campuses in sixteen years, and the cafeterias prepare Southern-style foods without the traditional ladles of fat. In fact, locally grown produce is included at lunch almost daily. In the Folsom Cordova Unified School District near Sacramento, California, Food Service Director Al Schieder, a former restaurant manager, replaced fast food with pasta, rice bowls, sushi, southwestern wraps, and oven-baked chicken—and turned a profit. Many schools have had the same experience: fix the food, and the money will follow.

In Sarasota, Florida, with its white sand beaches and miles and miles of orange groves full of sweet, juicy fruit ripe for the picking, school kids were drinking something called "Koolies," a drink made from sugar, water, and flavoring similar to Kool-Aid. French fries were classified as a vegetable under the school reimbursable meal program and sold as a vegetable almost daily. Nonetheless, the lunch program was bleeding money. That was twelve years ago, before Beverly Girard took the job as director of food and nutrition services for the Sarasota County schools. "People thought they were doing the right thing before I got there, but nutritionally it was a disaster and financially it was a disaster," Girard says.

A registered dietitian, Girard set about to fix the situation. She replaced the Koolies with a choice of milk or an eight-ounce glass of real orange juice. She encouraged the cafeteria staff to cut back on fries. In both cases, cafeteria managers said the kids would revolt. In fact, the students adjusted to both changes in a couple of days. At each step of the way, Girard sought consensus from the cafeteria managers at each school. She added more fresh fruits and vegetables, although they take more work to store and prepare than packaged food.

"The children are living in an area where produce is abundant, but in many cases it's not coming into the family home," she says. Girard secured funding to hire a nutrition educator from the University of Florida; she also consulted with the Sarasota Obesity Coalition on how to cut calories while preserving nutrition. Girard combined their professional advice with her own knowledge of kids' culinary limits. Her cafeterias do serve lower-fat cheese pizzas à la carte, for example, because the kids want it: she has to serve some kid-pleasers, she says, or most kids would leave campus and go to McDonald's.

Each time the food improved, the cafeterias made more money. Girard took the district's food service program from a $500,000 deficit to a $1.2 million surplus, and for her achievements she received a Director of the Year Award from the SNA. Yet despite these accolades, when Coca-Cola dangled a soda contract in front of her school board, she was powerless to stop it. "We have to fund programs. They're not lying that they need the money. It's just the wrong way to go about it," Girard says.

Girard and others fighting the obesity epidemic argue that without state or federal regulation of competitive foods, the nation's schools will keep making up the rules as they go along, to the detriment of kids' health. Until 1999, Florida had a strict competitive foods policy, but Governor Jeb Bush's administration rescinded it, saying that local control is best and that kids should be able to make up their own minds about what they eat in school. "We need to have a national (school) nutrition policy," says Girard. "Sweden has it. Norway has it. Other European nations have it. Why don't we?"

"So much of what we have to contend with is people who make decisions for us," she says. "In Sarasota County, I fought against the exclusive bottling contract. I almost laid my body down at a school board meeting. I almost quit. But I've gone back to them a million times and said one day this is going to be gone. And I'm going to say Hallelujah."

Sushi Is In, Burgers Are Out

Three thousand miles away, in the Folsom Cordova Unified School District near Sacramento, California, Al Schieder was as disturbed by the cafeteria food as Girard was. When Schieder, a native of Hungary, first arrived in the district in 1995, he was offended by the fact that poor kids had to line up for the federally subsidized school lunch, while the students who were better off could fill their plates with à la carte items—albeit unhealthy ones. "I realized we had two high schools. One was affluent. One was poor," Schieder says. He felt that the federal school lunch program's original intent in 1946—to provide nutritious food to all students—had failed. The à la carte foods, the greasy burgers, nachos, and doughnuts, he thought, were a nutritional horror. Along with unhealthy food and the basic inequality of the lunch line, the food service program was losing $200,000 a year.

A former chef and restaurant owner, Schieder thought like a businessperson. He went to the school district and asked for a loan to completely restructure his cafeterias. Then, in a bold move, he banished à la carte foods and soda. At the high school, he remodeled the cafeteria so that all the kids, regardless of income, could line up for the food they wanted. He put garden bars in every elementary school. He improved the breakfast program and brought it into the elementary school classrooms.

Now, high school students can choose from a selection of entrées, including pasta, rice bowls, sushi, southwestern wraps, oven-baked chicken, and a lower-fat cheese pizza. The entrées are rounded out with a piece of fresh fruit and a carton of milk. Every lunch is freshly made and geared toward the tastes of the lunchroom crowd. The skeptics said he would drive students out of the cafeteria and send the district into financial ruin. In fact his innovations turned a $200,000 deficit into a $300,000 profit by the 2001–2002 school year. Between 1995 and 2003, Schieder's food costs increased by about $500,000, and his labor costs increased slightly. But his sales more than doubled from $1.7 million to more than $3.5 million.

By banning à la carte items—and the fast food companies behind them—Schieder gained control over his lunchrooms. He uses his own recipes, and he no longer feels like a hypocrite since he stopped selling junk food in the same buildings where kids learn about nutrition.

"You know, we spend tremendous amounts of money to do nutritional education, and we have the knowledge, we have the computers; heavens, we have policies. And we have posters. We can wallpaper schools with (posters) about what nutrition is supposed to be. And yet we don't have the food," he says. Schieder contends that selling schoolchildren nutritious, appealing food at a reasonable price that brings a reasonable return on investment isn't rocket science and is within the reach of all food service directors. But the message isn't universally understood. "We're just standing there, and we don't know what to do. We operate like each individual little mom and pop store that's died in America. That's not business progress."

About one hundred miles to the west of Schieder's district, Aptos Middle School in San Francisco is another unassuming place for a school food revolution. Although it borders one of the city's wealthiest neighborhoods, surrounded by huge, stately homes with well-manicured lawns, Aptos is the city's most racially diverse school. The bulk of the children are African American, Chinese American, Latino, or Filipino; eleven percent are white. More than forty percent of the students come from families with incomes low enough to qualify for free or reduced-priced lunches. Here you won't find a gleaming dining room supplied by well-to-do parents. The school cafeteria is a plain brown and beige room where the students sit at worn brown tables that have seen generations of kids come and go. But now the food they eat each day is no longer hazardous to their health.

What happened at Aptos is a clear example of the difference just a few parents can make. After principal Ishibashi took her first horrifying foray into the Aptos cafeteria in 2002, she knew she had to do something. Tied up with countless other concerns, Ishibashi asked parent and active school volunteer Dana Woldow to figure out how to go about fixing the food that was sold à la carte in what's known as the school's Beanery. (Both the regular school lunch and the Beanery function out of the school cafeteria; the Beanery has a separate window.)

Woldow is a tall slender woman with short-cropped curly brown hair and chic glasses. She speaks at a fast clip and walks even faster, as if there were no time to waste in saving children from unhealthy foods. Until Ishibashi had asked her to help out, she had simply dismissed the fare at Aptos as "garbage" and had never allowed her kids eat it. Now, she knows where every gram of fat is buried.

"There is no one silver bullet," she says when it comes to fixing the problem of childhood obesity. "If it were that simple, we wouldn't be in a crisis. There are many ways of combating it, and you need all of them." Woldow believes that getting rid of soda and junk food and having solid nutrition education program are good places to start.

Woldow bypassed the nutrition department that had refused to give up chips and soda and went straight to the top. She got permission from School Superintendent Arlene Ackerman to start shaking things up at Aptos. She and parent Caroline Grannan put together a committee of parents and health professionals and, knowing how busy most parents are, tried to hold as few meetings as possible and communicate mostly by email. The group began by setting parameters for fat and sugar content by reading labels, although some foods destined for the chopping block were horrifyingly obvious—Slim Jims meat sticks, for example. The parents found the product contained not only several forms of sugar as well as sodium, but also bits of chicken tissue forced through a sieve. Slim Jims were out; so were chips, Hostess cakes, taco pockets, mega-colossal cheeseburgers (fifty-eight percent fat), hot links (seventy-seven percent fat), buffalo wings (sixty-one percent fat), giant pizzas, and french fries. Also off the menu were soda, Gatorade, and juices that weren't 100 percent fruit juice.

Woldow and Grannan next turned to the kids to find out what they wanted to eat. At the top of their lists were soup, Subway sandwiches, and sushi (after all, they are Californians). The school doesn't contract with brand-name fast food outlets, so the school's food service staff created their own healthy sandwiches with lean turkey or roast beef, lettuce, and tomato. They found a vendor who would make them sushi with no MSG. They brought in yogurt and fresh fruit and slimmed down the pizza to one slice with a salad on the side.

The new school food is a hit with its young consumers. The roast beef sandwich is fresh and tasty. The California roll sushi (no raw fish) wouldn't be out of place in any good sushi bar in San Francisco, and the soup is flavorful and brimming with vegetables. On one day, the most popular item with the kids was the spaghetti and meat sauce, although many of the sixth graders said they buy the soup often, and one said he's particularly fond of soup on days when he's had his braces tightened. His friend gave the new food a thumb's up: "I think it's good. It's really tasty." That day, they were serving the pizza with a side of fresh carrots and vacuum-packed peaches. The peaches got devoured; most of the carrots went into the trash. "You ever get your kids to try something new?" Woldow says, raising her hands in that helpless gesture that's universal to parents. "You have to put it in front of them ten, twelve, fifteen times before they eat it."

For two years now, the students at Aptos have been eating healthier food. At the end of the first year in 2003, the cafeteria turned a profit for the first time, and it has remained profitable. Students, it turned out, made better customers when they were served healthy food. Following the nutritional and financial success of Aptos, the San Francisco Unified School District began a district-wide "no empty calories" policy that went into effect in 2004, banning soda and junk food. And the Aptos experience brought other unexpected benefits. According to Woldow, there's less litter in the middle school's yard because there's less packaged food, and even better, fewer kids are ending up at the counselor's office with discipline problems after lunch.

"The most crowded time at the counselor's office used to be right after lunch when all the kids who got all sugared up at lunchtime would be experiencing their crash, and within fifteen minutes they got kicked out of class for some kind of bad behavior," Woldow says. "So it's people who have serious issues now who end up at the counseling office, not just a bunch of sugared-up kids." A relationship between sugar and hyperactivity isn't universally accepted, but Ishibashi confirmed that the post-lunch lines at the counseling office have thinned considerably since they changed the food.

The Aptos students' standardized test scores also improved after the Beanery foods were replaced. In the 2002–2003 school year, students at Aptos scored forty-five points higher on the Academic Performance Index (API) than on the previous tests, although the target was only for a seven-point improvement. Again, the parents can't prove a direct correlation, but numerous studies, including one at Tufts University, show that academic achievement is strengthened when students are well nourished.

To keep the kids motivated, Ishibashi gives raffle tickets for prizes to students she spots who are eating healthy food at lunch. "Now I find kids running up to me regularly, saying, 'Look, I'm eating this' or 'Look, I'm drinking water,' or having a salad, or whatever," the principal says. "They're making the cognitive connection."

Aptos has been so successful that Woldow can't keep up with her phone messages. She fields as many as five calls a day from Maine to Florida to Washington state from people asking for advice for their schools. In the final analysis, Woldow acknowledges that some districts that stop selling soda and junk food may lose money, but the cash they would have kept comes at a high price to kids. "If I went to the district, and I said that for $50,000 I can sell you a program that is going to calm your campuses, help your students pay attention in class, improve their health, and raise their test scores, they'd probably hire me to do that, wouldn't they?" she says. "They're getting it for free."

5

The Sedentary Bunch

IT TAKES A REAL FEAT OF ENGINEERING to make something in Las Vegas seem gaudy and excessive, but the architects who designed the Stratosphere Tower were up to the task. The tower glows with neon, a sort of Space Needle meets Liberace, and it's much taller than necessary. Take the ear-popping elevator ride to the observation deck, and you can practically see around the curvature of the Earth. When it was built in 1996, the adjoining Stratosphere casino and hotel gave Las Vegas another 86,000 square feet of gambling space and another 1,500 hotel rooms. The entire complex cost $550 million to build.

The Stratosphere dwarfs the other casinos on the Las Vegas strip with their dinky pyramids and puny Eiffel Tower replicas. It also casts a shadow over the most notorious neighborhood in town, a place where drive-through wedding chapels operate next door to motels offering hourly rates and free adult movies. In Las Vegas, "behind the Stratosphere" has become synonymous with "the wrong side of town." But for the kids at John S. Park-Edison elementary, it's home.

The Biggest Game in Vegas

Park-Edison looks like a typical Las Vegas elementary school with its crowded blacktop, largely ignored playground equipment, and squat rectangular buildings, including a couple of portable classrooms (trailers, actually) recently hauled in to handle the overflow of students. But Park-Edison isn't the average public school: like the tower that looms above it, it is actually part of a much larger corporation. In 2001, the Clark County School District handed over John S. Park and five other elementary schools to Edison Schools, Inc., a company that specializes in "rescuing" failing schools. Many parents protested the takeover—some even staged mock funerals—but the school district jumped at the chance to put financially and academically troubled schools in somebody else's hands.

Late on a May afternoon, twenty-eight of the most important stakeholders in this venture gather on the ragged grass-and-dirt field while the Stratosphere shimmers above them in near 100-degree heat. They are fifth graders waiting for their physical education class. Almost all of the kids are Hispanic, the children of Mexican immigrants who came to Las Vegas looking for good jobs and ended up washing dishes, cleaning rooms, or landscaping yards. A skinny blond girl stands out from the group, not only for her complexion but also for her long black pants and her cruelly thick turtleneck sweater. Most of the other kids are more sensibly dressed in shorts and T-shirts, but none of them looks like a fashion plate.

"Some of these kids wear the same shirt day after day," says Roy Leon, one of two PE teachers at Park-Edison. A few children will later walk home to one of the nicely kept middle-class houses surrounding the school, and a few will spend the night at a homeless shelter, but most live in a low-rent house or apartment building behind the Stratosphere.

These kids haven't had the best breaks in life, but they do have one thing that many kids in the country lack: an actual PE class taught by an actual PE teacher. Leon recently graduated from the University of Nevada at Las Vegas (where he played on the varsity soccer team), and now he's trying to get a bunch of kids hooked on exercise. He is wiry and boyish—if he put on a backwards baseball cap and baggy pants, he could walk down the halls of a high school without raising suspicion. His black hair is neatly cropped, and his smile is infectious. Even on a day when the sun beats down on the playing field, he expects everyone to have fun.

The kids crowd around Leon on the hot field. Some of the boys somehow manage an athletic swagger just standing in place. A couple of the girls try out their finely perfected looks of boredom. A few of the kids are thin and wispy, but several more are significantly overweight—if the class started picking teams for kickball or capture the flag, the competition for last-kid-picked would be intense. But on this day, nobody's going to be picked last. As far as Leon is concerned, no kid will *ever* get picked last. He is one of a new wave of PE teachers who's breaking the tradition of dodgeball and rope climbing. The "New PE," as it's known in the business, emphasizes physical activity above all else. The goal is to get as many kids moving as much as possible without making them miserable. Ideally, the kids will develop skills and discover passions that keep them moving for the rest of their lives.

After a round of stretches, the kids break into groups to play three-on-three keep-away. The patchy grass seems to get browner by the second under the blazing sun, but nobody complains about the heat, not even the girl in the turtleneck sweater. They complain about everything else, though. "He's cheating!" "She threw the ball too hard!" "Mr. Leon! I hurt my leg!" (It turns out to be one of those fifth-grader injuries that miraculously heals within seconds.) Throughout it all, the kids are running hard and their hearts are racing, and even the huffing-and-puffing heavy kids seem to be giving their best effort. A round Hispanic girl in a pink T-shirt runs after a loose ball as if she's being chased by a wild animal. And although they grumble, nearly every one of them says PE is his or her favorite class—at least when Mr. Leon is within earshot.

Leon tries to give his students more than a basic understanding of keep-away tactics. "I'm trying to teach them that exercise is healthy, and that the four major sports aren't the only form of exercise," he says. It's a tough sell. Leon grew up in a Hispanic household just down the street from Park-Edison, so he knows the neighborhood and the local culture well. He knows that most Hispanic parents don't put a strong emphasis on exercise for the sake of exercise, and he knows that most mothers and fathers are too worried about drugs and crime to let their kids loose in the neighborhood. "These kids don't go outside to play," he says. They don't have a decent recess, either, just five or ten minutes of free time after lunch. Aside from PE, they do have one outlet for their youthful energy and playfulness: when asked what they're going to do over the weekend, they shout, "play video games!" in near-unison. (These days, even kids who don't own enough clothes to get through the week have a Gameboy or a Playstation. The games are a big investment, but parents see them as a safe and easy way to keep kids occupied.)

Without PE, some of the fifth-graders at Park-Edison could go weeks or months without breathing hard. And without regular exercise, all of those calories from their typically fatty diets have only one place to go. Several of the kids in the class are already heavy, and a few are nearly too big to run, but they would all be worse off without their regular PE class, Leon says.

Nevada schools are some of the poorest in the country—each year, their spending per student falls about $2,000 short of the national average—and Leon worries that any further budget cuts could put PE squarely on the chopping block. After all, the state doesn't require a single minute of physical education for grades kindergarten through eighth, and gym supplies—not to mention gym teachers—cost money.

No Child Let Outside

For people of older generations, PE was a fact of life: if a kid went to school, he went to gym class. But that's no longer the case. According to the CDC, only about half of all schools require any sort of physical education for elementary school students, and only one-fourth to one-third require it for middle school kids. Many schools that do offer gym classes leave the task up to regular classroom teachers who don't have special training in physical education. Other school districts have cut PE costs by jamming enormous numbers of kids into each class. In California, the average gym class has more than forty students. In Las Vegas, a single class can have as many as seventy kids. In San Antonio, the number of kids *in one class* can reach triple digits. There's widespread fear that the situation will get even worse in the next few years. "PE teachers are very worried," says Charlene Burgeson, executive director of the National Association for Sport and Physical Education (NASPE). "They're worried about the kids not getting the physical education that they need, and they're worried about their jobs."

These fears have been realized with a vengeance in places like Hood River, Oregon, a small town surrounded by apple and pear orchards but famed for the windsurfing in the nearby Columbia River Gorge. In June 2004, the school superintendent laid off all five PE teachers in the elementary schools. Now classroom teachers who usually have little or no training in physical education are expected to add it to their list of teaching responsibilities.

The layoffs took place despite vocal and passionate opposition from parents and teachers alike. Cheryl Madsen, a first-grade teacher at the town's May Street Elementary School, says that there's no way she can give the children what the PE instructor provided. "I cannot do a quality program," she says. "How am I going to plan for this and get the equipment ready? I'm not going to be able to do it."

May Street Elementary PE teacher Stephanie Perkins was one of the casualties of the cutbacks. "I was shocked, especially when you can't turn on the radio without hearing about so many diseases that are coming up early with [overweight] children," she says. "I think the priorities are a mess."

Becky Kopecky, a registered nurse, is so upset about the PE cuts that she's considering sending her kids to private school. "My son had a fabulous PE teacher who focused on physical health and learning to work together with other kids," she says. "I have a nine-year-old son who has type 1 diabetes, and his blood sugars are so controlled when he's moving around. We can truly keep his insulin down and glucose under control with activity."

Her concerns are shared by Hood River pharmacist Kathleen Sanders, a certified diabetes educator who spoke out at the school board hearing. She noted that Hood River has a large percentage of Latinos, who have a disproportionately high incidence of type 2 diabetes. For that reason alone, she said, the town had a duty to do everything it could to keep those kids moving. "The community needs to do what it can to keep these kids away from this disease," she declared. "By eliminating PE positions, we're taking two steps away from the goal of preventing the adolescent onset of diabetes." She paused. "Maybe we're even contributing to the epidemic."

In Oregon and elsewhere, the threats to physical education are coming from many directions. For one thing, PE suffers from an image problem, Burgeson says. Many people in decision-making positions grew up during the dark ages of PE when classes were focused on athletics and competition instead of health and physical activity. Their most vivid memory of gym may be getting picked last on the softball team or dangling helplessly from the pull-up bar.

One PE teacher, in fact, offers a novel theory as to why so few principals support physical education. "Those kids who couldn't climb ropes, those kids who couldn't do a push-up, you know what happened to them?" half-jokes Susan Kogut, a former physical education teacher, now a lecturer at the University of Maryland. "They grew up to be principals."

Even if PE teachers can improve their profession's image, many will still face the obstacle of chronic underfunding. Schools across the country are struggling to pay for textbooks and teacher's aides, and many are falling behind state or national standards for academics. As a result, they have to squeeze more instruction into each day with fewer and fewer resources, making "extra" subjects such as art, music, and PE suddenly seem expendable.

The pressure to cut PE and other programs grew especially intense with the passage of the No Child Left Behind Act of 2001, says William Dietz of the CDC. The act financially penalizes schools whose students do not score well in standardized reading and math tests. Many schools have responded to this threat by cutting back on "nonessentials" and allotting more and more time and resources to helping kids drill for the tests. The full impact of No Child Left Behind has yet to be felt, but the early results are worrisome. "We're collecting stories from around the country," says Daniel Kaufman, a spokesperson for the National Education Association. "Recess is cut back, and there isn't as much time for PE."

Ironically, the act ends up placing a particularly large burden on disadvantaged kids at poor schools, the very children who supposedly weren't going to get "left behind." Kids at poor urban schools are especially likely to fall short on standardized tests, Kaufman says, which means their schools are the most likely to cut PE. Low-income children are also especially likely to be overweight, he says, making the loss of PE that much more painful.

As elementary schools try to jam more instruction into each day, even recess is becoming a luxury. According to the CDC, roughly thirty percent of elementary schools nationwide don't offer regular recess. "A lot of kids get a few minutes after lunch, and that's it," says Melinda Sothern, PhD, an exercise physiologist with Louisiana State University and lead author of the popular book *Trim Kids*. In many ways, the demise of recess is even more troubling than cuts in physical education, she says. Recess gives kids a chance to burn energy on their own terms, and those games of four square or catch are often more physically demanding than PE classes.

Any principal or administrator who tries to improve standardized test scores by cutting back on PE and recess probably deserves a failing grade in basic biology. "I think it's a very shortsighted approach," says James Sallis, a professor of psychology at San Diego State University and a nationally recognized expert on promoting physical activity in children. "It assumes that kids can sit still for six hours and absorb the information like robots. They're kids, and they need breaks. It's almost inhuman." There's no evidence that hours of cramming actually improves scores, Sallis says, but there's plenty of evidence that students are more likely to gain weight if they're stuck at their desks all day. "They're not fixing the problem that they're trying to fix, and they're causing another problem."

When kids miss out on PE and recess, they lose their best chance for physical activity. In theory, they could make up for those hours stuck inside by running rampant through the neighborhood after school. Unfortunately, they don't do that. A study of third and fourth graders published in 2000 found that kids were actually *less* active after school on days when they don't have PE.

The War for Four Square

PE classes and recess are under siege at the worst possible time, says former Surgeon General David Satcher, now the director of the National Center for Primary Care at the Morehouse School of Medicine and a founder of Action for Healthy Kids. The epidemic of childhood obesity will only grow worse if kids aren't given more opportunities to be physically active, he says. For this reason, Satcher was happy to take time from his packed schedule to testify in March 2004 on behalf of a recess bill before the Georgia state legislature. "Believe it or not, this bill was to require the schools to give at least fifteen minutes a day of recess," Satcher says. "Can you imagine what's gone on over the years in schools when children don't even get recess, let alone [physical education]?"

Unfortunately, even a surgeon general has limits to his influence: Satcher failed to persuade Georgia legislators. Instead of agreeing that children should be able to run around and play outside for fifteen minutes each day, lawmakers passed a watered-down version of the bill, removing the mandate for recess altogether. Georgia school boards "may" require recess, but they can just as easily decide to keep children inside for the entire school day without a break. It's no mere coincidence that when researchers at the University of Baltimore rated states on their efforts to control obesity, the state of Georgia got a failing grade.

What happened to the school recess bill in Georgia is a snapshot of what is happening in communities across the country. Even so, the bill's sponsor was stunned that she couldn't get fifteen minutes of daily recess to pass muster with her colleagues. "People would say this is kind of a no-brainer—of course kids should get recess," says Representative Sally Harrell, a Democrat from DeKalb County.

School superintendents argued that children don't really need time to run and play at school. "I thought that's what home is for," says Ruel Parker, superintendent of the Rockdale County schools. Parker insists that students aren't sedentary at the schools he oversees, noting that they have some type of PE. Also, they walk from class to class. "Granted, it's not calisthenics, but it's activity," he says.

Walking between classes is no substitute for recess, says middle school teacher Deanna Ryan. Many of the students in Georgia's middle schools students are trapped indoors all day, she says. "In Georgia, there are hardly any middle schools that provide any significant break. They give them between two and five minutes between classes."

Five-year-olds are also kept on a tight leash. Olga Jarrett, a Georgia State University professor of early child development who has studied recess throughout the state, says that twenty-five percent of kindergartners in Georgia—who are mandated to have a full day of school—do not have daily recess. Even in the lunchroom, where laughing and poking and squirming are as traditional as mystery meat and hairnets, kids are strictly controlled. "They have twenty minutes of lunch, and often they can't talk during [that time]," says Jarrett. "A red light goes on if they're talking too loud."

Supporters of recess, which included parents, teachers, and the local chapter of the American Academy of Pediatrics, were outraged by the opposition to the bill, and they flooded the state capitol with letters. "Who can maintain focus and good humor for seven hours?" wrote Michelle Gregg, a mother who volunteers in her child's first-grade classroom. "No adult can achieve that, much less seven-year-olds. Behavioral problems and 'acting out' will become out of control, and there will be no focus on academics as a result. It's already a feat to get seventeen grouchy, cooped-up kids to get their bodies under control long enough to focus on anything."

Gregg went on to say that, "If we take away their outdoor play time, we are in effect teaching them that it's okay to sit around all day and become fat and unhealthy. School playgrounds are some of the few places for city children to go outside and refresh themselves safely. We cannot jeopardize their health any more than we already are."

Another parent, Anne Torrez of Gwinnett County, was so upset by the prospect of her child losing recess that she collected 400 signatures in protest. She then offered to monitor recess at her children's school as a volunteer, but was told that she lacked the certification to do so. "My nine-year-old daughter only had recess five times this last year," said Torrez. "This is just ridiculous. I told my husband that if our kids can't have recess, I want to move."

Torrez's daughter, Cydney, adds that what she likes best about recess is climbing on the monkey bars, running, and playing with her friends. "When I don't have recess I just feel like I want to jump out of my seat and run around the classroom," she says. "But when I do have it, it helps me sit down and work better. And all the other kids seem less wild when we do have recess."

Georgia's Lieutenant Governor Mark Taylor echoed parents' and students' concerns and supported the recess bill, especially for elementary school children. "Overall, people would like to see kids enjoy the school day, have interactions with classmates, burn off energy and excess calories, which might help them become healthier adults," he said.

Given that logic, why would school boards and superintendents be so virulently opposed to recess? Taylor attributes it to the state's—and the nation's—emphasis on school testing, including that required by the No Child Left Behind Act. "The rallying cry for the last twenty-five years is accountability. Get those test scores up. Why can't Johnny read?" he said. "This has led state, federal and local school systems to insist on almost every minute of the school day to be accounted for . . . and recess and physical education have been losers."

The nation's parents apparently agree with Taylor and the angry mothers in Georgia. A national survey conducted by the American Obesity Association found that the majority of parents in the United States (seventy-eight percent) believe that physical education or recess "should not be reduced or replaced with academic classes."

Despite the failure of mandatory recess in Georgia, Satcher hasn't stopped his crusade. "We have to close the gap between our science and our policies," he recently told a packed audience of PE teachers at a conference in New Orleans. "I can think of no better example of that than the underfunding of physical education. We spend $120 billion a year treating the consequences of obesity, but schools say they can't afford PE classes." The standing ovation lasted for a solid minute.

Sitting Still

Satcher has warned that the youngest generation is already the most sedentary in history. By the latest estimates, half of all kids (and seventy percent of all adults) get less than thirty minutes of exercise each day. Thirty minutes isn't an especially lofty standard, especially when you consider that the NASPE believes kids should get at least one hour of moderate to vigorous activity every day. One hour a day (or, expressed in kid time, just two episodes of the *Power Puff Girls*): That's all it would take for kids to reap all of the benefits of exercise. Their hearts would be stronger, and, not incidentally, they'd be much less likely to put on weight. They'd feel more capable, confident, and energetic. They'd also feel happier. Several studies have found that exercise relieves the symptoms of anxiety and depression, all-too-common problems for today's kids. "The mental health benefits may be the biggest motivation for kids to exercise," Sallis says.

And contrary to what many school administrators seem to think, physical activity doesn't come with a downside. Many studies have shown that grades and test scores don't drop at all when schools allot more time to physical education and recess, Sallis says. In fact, he has conducted studies suggesting that top-quality physical instruction can actually help students score better in other subjects. Some administrators have already taken that lesson to heart. "PE is the place where you can truly change kids' lives. Most kids have never truly felt what success is. You set a goal, you make a commitment, and you have the work ethic to see it through," says Alice Mehaffey, principal of Selma

Middle School in Selma, Indiana. "If your kids feel good, if they are more active, they are going to do better in every other classroom."

Kids have plenty of reasons to get moving, but many end up sitting still. You can see it in Las Vegas, and you can see it in Selma, the anti-Vegas. Selma Middle School is flanked by two cornfields and a pig farm, and the closest thing to the Stratosphere is a nearby grain silo. The town has a few baseball diamonds, but kids aren't exactly awash in exercise options, says Tammy Brant, the PE teacher at the middle school. "These kids can't even walk to school because they'd get run over by a tractor," she says.

Brant grew up in Selma, and she says the town itself hasn't changed much over the years. The people, however, are another story. When Brant was young, most of the local kids had two parents: one who worked, and one who stayed at home. Today, she says, far more kids live in single-parent households, and almost all parents have full-time jobs. As a result, parents don't have a lot of time to play outside with their kids or drive them to nearby Muncie, a relatively large town with a skate park and a swimming pool. And when parents are at work, they usually don't want their kids roaming the neighborhood unsupervised. Kids are spending more and more time inside, and Brant believes it's starting to show on their waistlines. "The obesity rate is getting high here because kids don't have the opportunity to do things," she says. The school nurse recently put the kids on the scales and found that nearly one in three was overweight.

In Las Vegas, Selma, and everywhere else, kids are finding it easy to be inactive. They live in a world of convenience, a world where practically all new cars come with power windows and handheld CD players come with remote controls. Television, computers, and video games all constantly vie for a bigger and bigger share of a kid's free time. Microsoft used a $500 million marketing campaign to launch its Xbox gaming system, which is exactly $500 million more than the advertising budget for freeze tag. Meanwhile, more parents are driving children to their activities and discouraging unsupervised play. The general public is rightly worried about the epidemic of childhood obesity, but the twin epidemic of inactivity often goes ignored, says Judy Young, PhD, executive director of NASPE. "Many people think that young kids are just naturally active," she says. "If anything, they're focused on getting kids to settle down."

Health experts worry that kids are settling into a sedentary lifestyle that will jeopardize their health for decades to come. As recently reported in the *Journal of Applied Physiology*, a lack of exercise—by itself—raises the lifetime risk of at least seventeen chronic conditions, including type 2 diabetes, heart disease, arthritis, several types of cancer, and of course obesity. Inactivity is especially hard on young bodies. Children who don't exercise may develop complications such as high blood pressure, high cholesterol, or insulin resistance (a precursor to diabetes) before they're old enough to get a driver's license.

It's worth stressing that a lack of exercise *alone* can be enough to put a target on a child's back. Whether a child is fat or thin, sitting on her bottom all day is unquestionably hazardous to her health; in fact, some experts believe that inactivity is even more dangerous than obesity. On the positive side, regular activity is good for a kid no matter what his size. "If we could just get our kids more active and not worry about anything else, we'd lower the risk of many diseases," says Timothy Lohman, a professor of physiology at the University of Arizona. If a heavy kid starts riding her bike to school every day or takes up the daily dog-walking duties, her health will likely improve even if she never sheds a pound.

It's a good thing exercise has benefits beyond weight loss because heavy kids or adults very rarely manage to walk, run, swim, or bike their way to a slimmer body. As William Dietz recently told a conference of PE teachers, "It's widely thought that people who are more active will lose weight, but it's not so. Physical activity adds little to weight reduction." Exercise alone may not burn off the pounds, but it can definitely help keep those pounds from piling up in the first place. No matter what a person's age, regular physical activity can help prevent the buildup of fat, making exercise a crucial component of any successful long-term weight loss program.

The workouts don't need to be brutal either. As reported in the *Archives of Pediatrics and Adolescent Medicine*, a ninety-pound child can prevent gaining over one pound of fat each year by simply walking an extra ten minutes each day. That same child could avoid over four pounds of fat if she simply turned off the TV for an hour each day and spent at least thirty of those minutes walking. The other thirty minutes would give her an excellent chance to read or catch up on homework.

The Failures of "Fitness"

Physical activity is a worthy goal in itself, but many parents and teachers have something else in mind: extreme fitness. Where other people see kids, they see future aerobic instructors. They expect all kids to be able to fire off pull-ups or fly around the track, and they push school districts and states to establish elaborate "fitness standards." Even without official standards, many gym teachers take it upon themselves to turn all of their students into little athletes, including the heavy ones. But there's a problem: the emphasis on fitness doesn't make kids fit. It just makes them miserable.

"We've pushed fitness for years and years and years in America, but things haven't improved," says Robert Pangrazi, professor of exercise science and physical education at Arizona State University and author of *Dynamic Physical Education for Secondary School Students*, perhaps the closest thing gym teachers have to a Bible. "The fitness push failed obese kids. It was never designed for the kids who need it the most.

If you add 20 pounds of sand to your back, you're not going to be able to run as fast or perform as well. Fitness scores are going down because kids are getting fatter."

In fact, the most heralded effort to improve fitness in recent years quickly became a lesson in demoralization. "We were going to motivate kids with the Presidential Award for Fitness," Pangrazi says. "To get the award, kids had to finish in the 85th percentile in several categories. Only one-tenth of one percent qualified, and 99.9 percent of kids were labeled as losers."

Bodies in Motion

The real goal for parents and teachers should be keeping kids physically active, Pangrazi says, not making them strong, fast, or skinny. Anything a kid can do to get moving and burn calories is a positive step. Cleaning her room, throwing a Frisbee, raking leaves, riding a skateboard, walking beside the cart at the grocery store . . . it all counts, and it all adds up. Staying physically active doesn't take a monumental effort, but around the country and around the world, many kids continue to idle their way toward disaster.

The sad irony is that little people are built for motion. As Sallis explains it, the brains of young kids are loaded with special pleasure receptors that respond directly to physical activity. Whenever a youngster runs or jumps or hops or dances, he's getting more than just exercise—he's getting a rush. Even though kids get deep pleasure from exercise, it's still possible to slow them down. It just takes a lot of effort. "We work very hard to keep kids still," Sallis says. "They hear the mantra to sit down and be still a million times. We structure their days to keep them still. We're trying to make them fit into adult patterns, and we're fighting biology the whole time."

The fight to keep kids motionless becomes easier and easier over time, Sallis says. It's a fact of nature that physical activity declines with age.

It happens to monkeys. It happens to fish. It happens to fruit flies. And it definitely happens to humans. From preschool on, kids became gradually less and less active, partly because the brain's pleasure centers that reward kids for exercise gradually start disappearing. Seven-year-olds run around less than five-year-olds, and by the time kids become teenagers, their brains are seeking out other types of satisfaction.

While all teenagers tend to shift into low gear, girls are especially likely to take inactivity to extremes. Researchers at the University of Pittsburgh recently tracked the physical activity levels of more than 2,000 girls for ten years starting at ages nine and ten. The girls gradually slowed down as they got older, and by the time they hit their teens, their activity levels fell dramatically. At ages fourteen to fifteen, practically all of the girls had at least some activity. But at ages sixteen to seventeen, an incredible fifty-five percent of African American girls and thirty percent of white girls scored a complete zero on the activity scale.

Rides and Remote Controls

So what are kids doing with their time? For one thing, the average American child watches three to four-and-a-half hours of television every day. It would be bad enough if TV simply exposed kids to reality dating shows and Carrot Top, but it also seems to help make them fat. A 1996 study by researchers at Harvard and the CDC found that kids who watched more than five hours of TV each day were more than four times more likely than other kids to be overweight.

"Watching a lot of TV is probably the biggest single risk factor for becoming overweight," Sallis says. Most people see a straightforward story: Kid spends all day on the couch watching the tube, kid gains weight. But, surprisingly, TV doesn't necessarily make kids inert. A 2003 study by researchers at the University of Montreal found that high schoolers who watch a lot of TV get just as much overall physical activity as those who watch very little. (It turned out that the teens who weren't watching *American Idol* or *The Real World* were spending their time reading or playing on the computer or talking to friends on the phone.)

When TV addicts put on weight, it may have more to do with what they put in their mouths than how much time they spend on the couch, Sallis says. Kids who watch TV typically don't just sit and stare—they sit, stare, and snack. And what do they snack on? The constant barrage of ads for fatty or sugary food on children's television pretty much answers the question. For many kids, the TV habit becomes hopelessly intertwined with the junk food habit, and both end up working together to make them fat.

Computers and video games are also helping to keep kids glued to screens. Between Madden NFL, MVP Baseball, and Backyard Soccer, who has time for a real game? Nobody knows how much computers and video games might be contributing to the obesity epidemic, but there's cause for concern. A study of 2,318 kids ages nine to twelve living in a low-income area of Montreal found that playing video games every day more than doubled the risk of gaining excess weight over the next year.

The biggest activity drain of all may be the family car. Kids today catch rides just about everywhere they go, especially on their way to and from school. The youngest generation is sick of hearing stories about how their parents and grandparents had to walk miles to school every day—uphill both ways, in waist-deep snow every month of the year—but there's some truth in those tall tales. Marya Morris, head of the Planning and Designing the Physically Active Community project at the American Planning Association, regularly speaks to groups about healthy communities, and she often asks two simple questions. First she asks, "How many of you walked or rode a bike to school when you were children?" Almost everyone raises a hand. Then she asks, "How many of you drive your kids to school?" Again, almost everyone raises a hand.

On any given weekday morning, the streets in front of a typical elementary school are jammed with double-parked station wagons and minivans. According to the CDC, today's kids ride their bikes or walk between school and home only about thirteen percent of the time. In 2000, a survey in Marin County north of San Francisco found that one in four cars in morning traffic was taking kids to school.

Why are so many parents taking up chauffeur duty? For one thing, parents today are more worried than ever about their children's safety. The fear is often justified: some streets are too busy for kids to cross, and some neighborhoods aren't safe for walking. In other cases, the danger is more perception than reality. Violent crimes against children aren't really more common today than they were twenty years ago, but media reports of a few high-profile cases have put many parents on edge.

Even if kids had their parents' blessing to walk to school, for many it would be an all-day trip. A growing number of kids go to "magnet schools" that enroll kids based on their interests, not their neighborhood. Others live in sprawling suburbs where the nearest school of any sort is a long, long, long, bike ride away. Schools in growing suburbs are typically built on the biggest, cheapest lots available, Morris says, which often means a field out in the country far away from the booming neighborhood.

Suburban sprawl puts up other barriers to exercise, Morris says. In some places, neighborhoods stretch seamlessly for miles with no parks or stores to break the monotony. If kids did want to ride their bikes or go for a walk, they'd be hard-pressed to find a destination. The main streets are generally designed to accommodate the maximum amount of traffic moving at the greatest possible speeds—not exactly the ideal scenario for a 10-speed Schwinn. Traffic moves more slowly on the side streets, but it still isn't easy for nonmotorists to get around. For one thing, the side streets don't always connect easily to each other, she says. A walk that should only take fifteen minutes could easily take thirty minutes—even more if a person happens to get lost in a maze of cul-de-sacs and dead ends. Aerial shots of the subdivisions in San Antonio tell the story, says Joan Miller. "The school may be over your back fence, but you live in a cul-de-sac, so you have to drive out and go way around to get out of your neighborhood and onto the thoroughfare to get to the school," Miller says.

In short, the "good life" of suburbia comes with a price. A 2003 study by researchers at the CDC estimated that suburban life adds an average of six pounds to every adult.

"Left in the Dust"

Whether kids are stuck in the suburbs or camped out on a couch, inactivity sets an insidious cycle in motion. Inactive kids are more likely to gain weight, and as they get larger, it becomes harder and harder to get in the game. Not only are heavy children often crippled by embarrassment and self-consciousness, they also lack the strength and stamina to keep up with other kids. For Molly Markus of Billings, Montana, a few extra pounds made the difference between running with her friends and, in her words, "getting left in the dust." As a ten-year-old fifth grader, she stood 4'9", weighed 145 pounds, and always brought up the rear during gym class. "Everybody was skinnier than me, and when we ran, they would have to slow down and wait," she says. "But when someone is different, they don't want to wait."

PE teachers have deep sympathy for heavy kids, Pangrazi says. The teachers hear the teases and taunts, and they know as well as anyone how much overweight kids struggle to keep up. Unfortunately, this sympathy often brings out the drill sergeant in well-meaning teachers, he says. Too many teachers act as if a few more laps around the track will solve all of a heavy kid's problems. "We think we need to run the hell out of them and get them skinny, and it just doesn't work," he

Carmen's story: "It's better if you lead yourself"

For sixteen-year-old Carmen Miranda, who lives in a low-income neighborhood in Richmond, California, extra weight isn't such a big crisis. (Carmen's mother didn't name her for the famous singer, although, annoyingly, anyone over thirty usually asks how she got her name.) Of Latina heritage, the dark-eyed teenager, like her friends, favors black, baggy clothes. She's a self-possessed kid whose attitude is clear: you can like her or not. It doesn't matter to her. Carmen doesn't have diabetes or any other major health problem. And because she's a bit older than Cheryl (see Chapter 1), she has more control over her time and her life. Carmen has decided to use that time to lose weight and get fit. At 5'7", she weighs 246 pounds.

Already heavy in the fourth grade, Carmen had trouble keeping up in PE, but she wasn't anxious to do anything about it. "I would just go home, do like nothing. Just watch TV," Carmen says. "I don't know. I just got lazy. I didn't care, and if I'm fat, I'm fat."

In eighth-grade gym class, she went from lagging behind to constantly gasping for air. "I couldn't run the whole thing. And like I didn't feel good," she says. "I saw other people and I was like, I wish I was like them. But then I didn't care as much. So I kept on eating." On days when the school weighed students, Carmen stayed home. She was too embarrassed to have someone put her on a scale.

This year, Carmen decided she had a reason to lose weight: she wants to get down to 170 pounds so she can fit into a slinky prom dress. She envisions herself with the curvy body shape of Drew Barrymore—after all, she already has Barrymore's feisty attitude. Her weight loss and exercise regimen is strictly for herself—not for her friends or her family. "It's for me because I'm the one who's carrying it. I'm the one who's trying to lose it. It's not hard if you really put your mind to it. But it's hard if you just, like, let other people lead you. It's better if you lead yourself."

Her mom had been encouraging her to lose weight, although not necessarily in a way that Carmen thinks is helpful. "My mom is like, 'You ate too much today. That's why you're going to get fatter.' I'm like, okay, Mom, it doesn't help." Carmen says her mother also cooks a lot of fried food that she's trying to avoid.

Until this year, a typical lunch for her in school was a soda and Flaming Hot Cheetos or a McDonald's cheeseburger followed by a Snickers bar. The teenager favors fast food, candy, and soda, but she began experimenting by cutting back on those foods. She lost five pounds in the first month, but then she lapsed back into eating junk food again.

After that, she made a deal with her skinny friend Roberto. She told him, "If I eat something fattening, just knock it out of my hand. I don't care if I get mad. I'll be happy in an hour." So Roberto did, literally. He'd smack the chips or the

➔

cheeseburger right out of her hand. Roberto's energetic response to her fatty food choices has made her more likely to eat a regular sandwich for lunch than a Big Mac.

Roberto and her other friend Rita have also teamed up as Carmen's exercise squad. Around the same time she started changing her diet, Carmen's father agreed to buy her a membership to Curves, a chain of women's fitness centers. Sometimes Carmen and her friends walk two miles to Curves, Carmen exercises for thirty minutes, then they all walk home together. Over three months, Carmen lost fifteen pounds. Her exercise squad has been crucial to her efforts to be thinner and fitter. "I really listen to them," Carmen says. "I'm like with them every day and talking to them on the phone and everything."

She's also had some help from an after-school program where she learned how to read food labels. "If you want something really bad, and you read the ingredients and the first four words are sugar, don't get it." •

says. "Running may be the worst thing you can do for overweight kids. I still see the mile run used on these poor little obese kids. It's a painful experience, and they learn to hate it." Average-weight kids reach about sixty percent of their breathing capacity while running a mile, he says, but overweight kids are burning their lungs at about ninety percent of their capacity. That's the difference between a good workout and complete agony.

Turning on the Motors

Of course, physical activity doesn't have to be painful, even for the heaviest kids. It doesn't have to be boring either. Kids of all sizes and ages can find a type of exercise that feeds their passions while working their bodies. Even teenagers can get moving with the right encouragement, Sallis says. Every kid has a motor—some just need a little help finding the switch.

Before the first bell rings at Fay Herron Elementary School in North Las Vegas, a throng of kids gathers in the central courtyard. A few are jumping rope or playing four square, but most are just standing shoulder to shoulder and back to back with their classmates. With 1,400 students, Fay Herron is the largest elementary school in Nevada. A generation ago, it was a mostly African American school; today, at least ninety percent of the kids are Hispanic. The kids live within walking distance of the Silver Nugget Casino—the giant marquee in front reads, "Jackpots paid in five minutes or less or you eat free"—but this is not the Las Vegas of the TV commercials or the tourist brochures. Fay Herron sits in a neighborhood of small houses and trailers where most of the lawns have gone feral. Because the school doesn't have a single school bus, almost all of the kids walk to school every day, a trek that takes them past palm trees, graffiti, and junked-out cars.

Like Park-Edison on the other side of town, Fay Herron has something that many wealthier schools lack: a real PE teacher. In fact, they have a teacher who looks like he came straight from the gym teacher textbook. Jurgen Kraehmer is lean and muscular. Much of his reddish hair has fallen out, but you can still sense the presence of a military-style haircut. He has a second-degree black belt in karate, he has run a couple of marathons, and he has an air that commands respect. And he'd better: his typical class has sixty to seventy students, or about thirty to forty kids past the point of sanity. He has an assistant, but she can only deal with one unruly kid at a time. Without respect, there would be chaos.

Kraehmer may look "old school," but, like Leon, he's a strong proponent of the New PE. His kids won't spend half the class standing in line or waiting their turn, they're never "out," and nobody will be the last one picked for a team. Most of all, they won't get discouraged with themselves. "I can't change their fitness, so I have to change their attitude," Kraehmer says. The ultimate goal is to make them more active in their everyday lives. He tries to give them the basic skills they'll need to enjoy games and sports, and he gives them fitness goals that are well within reach. He also tries to help them see the big picture, but it's a challenge. "Being healthy and fit? They can't see that far," he says. "They're worried about having clothes and food for the next day. I just tell them you need to be healthy so you're not stuck in bed."

On some days, the kids will crowd into a carpeted PE room for some hip-hop line dancing. On other days, they'll work out to Tae Bo tapes playing on the television monitors that Kraehmer has purchased with the help of federal grants. The walls of the room are lined with brand new weights and medicine balls, also purchased with grants secured by Kraehmer. (Before he bought the weights, he showed the kids how they could strengthen muscles with plastic bottles filled with sand.) Today, they will play kickball, but not old-style nine-on-nine kickball or, perish the thought, twenty-five on twenty-five. This is the New PE, and nobody will be planted in left field.

On a hot spring day (the only kind of spring day there is in North Las Vegas), fifty or so third graders play a quick game of red light/green light on their way to the playing field. Once they reach the grass, they all drop for ten pushups while Kraehmer takes roll. (They often do their pushups on the blacktop, but today the asphalt is too hot to touch.) At least ten kids are absent, making the field a bit less crowded than usual. After the pushups, they take a quarter-mile run around the playing field. Well, some of them take a run. About ten of the kids settle for a slow jog, and a few heavy kids near the back never break out of a walk. Kraehmer urges them to pick up the pace, but they don't seem to hear him.

Kraehmer is especially concerned about his overweight students. He knows that many of them already have mixed feelings about exercise, and they may be just one humiliation away from quitting entirely. "I expect them to do what they can," Kraehmer says. "I don't want them to hate it, but I want them to do it. A lot of kids have given up on themselves."

After the run, the kids line up for a quick drink of water before getting to the main event. Kraehmer rolls out a cart stuffed with kick-balls and breaks the kids up into seventeen groups of three. Each group will play their own game: one will be the pitcher, one will be the fielder, and one will be the kicker. Each group gets two small mats for bases. They find a spot on the field, put down their bases, and the game starts. Before long, kids and balls are flying around the field like drops of water on a hot griddle. Even the heaviest kids seem to find another gear when they're trying to run down a ball or beat a throw to home plate. Nobody seems to be keeping any sort of score, and that's fine with Kraehmer. This isn't about winning and losing—it's about keeping kids moving.

$1 Billion in the Right Direction

No Child Left Behind not withstanding, the U.S. government has also made a commitment to get kids moving. For example, the CDC recently launched VERB, a large TV ad campaign designed to encourage nine- to thirteen-year-olds to become more physically active. Dietz calls it the first such campaign ever to get adequate funding, and it's getting results. "Kids exposed to the campaign have become demonstrably more active, especially girls and inner-city kids," he says.

One recent TV ad features a kid playing "H-O-R-S-E" on a decrepit city basketball court while an ESPN announcer provides commentary. The commercial ends with the message "Horse. It's a sport." Different ads are tailored to certain audiences. There are ads aimed at girls, ads aimed at boys, and ads targeting different racial and ethnic communities. A particularly striking ad aimed at Native Americans shows a boy struggling to encourage his friends to go outside and play. Frustrated, he calls his dad at the power station. Reluctantly, the dad agrees to turn the power off for one hour. The TVs and the computers go quiet, and kids stream outside.

The government is also making a major push to encourage kids to walk to school. The walk to and from school gives kids a chance to gather their thoughts and stretch out their legs, and it also may prove to be one of the most effective remedies for the epidemic of childhood obesity, Sallis says. He estimates that the average child would burn about fifty calories walking a half-mile to school. Walking twice a day for 180 school days would burn about 18,000 calories per year—enough to prevent a weight gain of five pounds. As concern about childhood obesity grows, several cities have recently received federal grants to encourage kids to walk or bike to school. Among other things, the money goes to create bike lanes, install sidewalks, put up stoplights, and build pedestrian bridges. Many schools—including 175 in the Chicago area alone—have started "walking school bus" programs where parents walk to school with groups of neighborhood kids.

Communities that have already invested in Safe Routes to School (SR2S) projects have seen substantial returns. California's Marin County—where one in four cars in the morning traffic was taking kids to school—started an ambitious SR2S program in 2000. In addition to building new bike lanes and sidewalks, the county enhanced crosswalks, put up new signs, and changed the timing on street lights to make it easier for pedestrians to cross. In just two years, the number of kids who walked or rode their bikes to school increased by eighty percent.

These small steps here and there are about to expand into a nationwide movement. SR2S programs will soon start up in all fifty states thanks to strong support from various politicians—notably James Oberstar (D–MN) in the U.S. House and Jim Jeffords (I–VT) in the Senate—and the promise of substantial federal funds. In its version of the 2004 transportation bill, the U.S. House of Representatives proposed spending $1 billion over six years for SR2S projects. The senate version of the bill would allot $420 million over six years.

One billion dollars over six years may seem like a lot of money, but even that won't come close to giving every kid a decent route to school, says Martha Roskowski, the campaign manager for American Bikes, a Washington, D.C., advocacy group working to secure funding for SR2S programs. "It's a small investment compared to what the need is," she says. Roskowski estimates that it costs about $250,000 to build a mile of sidewalk. When that billion dollars gets spread across the country, a state such as Oklahoma—with its 1,800 public schools—would be able to afford nine miles of new sidewalks, she says.

Still, that billion dollars could make a huge difference if it stimulates cities, counties, and states to chip in some of their own money. "There's nothing like federal grant money to get people excited about a project," she says. The highway industry tends to grumble about any project that takes money away from road construction, but, overall, the SR2S movement has run into very few obstacles. "It's amazing how little opposition the project has gotten," she says. "My guess is that there are a few politicians who are unhappy, but they don't want to say any thing about it. They'd get pilloried. Who could be against kids walking and biking safely to school?"

Here's another relevant question for these times: who could be against active, livable neighborhoods? For the past fifteen years or so, urban planners have been looking for better ways to build communities. Instead of signing off on yet another sprawling expanse of houses, they are encouraging more compact communities with mixtures of houses, businesses, and parks. This movement—often called "smart growth"—was originally intended to promote vibrant communities, ease traffic congestion, and protect farmland and other open space. Eventually, people realized that "smart" communities also promoted physical activity, putting urban planners and public health experts on the same team.

"The obesity epidemic has created a whole new level of interest in the design of our communities," Roskowski says. "There is a leading edge of the health community that is looking for ways to help people keep from getting sick. They pretty quickly come to the conclusion that people need to bike and walk more."

The threat of obesity has reenergized efforts to promote smart growth, Morris says. "Planning is very politicized," she says. "The fact that we have people with MDs backing us up makes a huge difference." Morris and others still run into plenty of skepticism, but now they have research in their corner: "A lot of people kind of roll their eyes when I talk about this. They say, 'Are you telling me that wide streets make people fat?' They're more likely to believe it now." Indeed, many experts now believe that urban planning may play an important role in the country's battle against obesity. Government agencies such as the CDC and charitable organizations such as the Robert Wood Johnson Foundation are wholeheartedly putting their expertise and money into this new arena. The RWJF alone has committed more than $70 million to promoting smart growth and active communities.

Although many cities and towns have made small changes to increase physical activity—a bike lane here, a walking path there—one new town is taking "smart growth" to extremes. Stapleton, Colorado, a community growing on the plains east of Denver, fits the dictionary definition of a suburb, but it certainly doesn't fit the mold. Stapleton will have a network of parks and trails that connect people to grocery stores, restaurants, and other destinations. Planners expect to attract 30,000 residents in the next decade. This small town will also be a major experiment in a state that already boasts the nation's thinnest residents. If community design really does matter, the families of Stapleton should turn out to be slimmer and healthier than typical suburbanites.

Molly Markus doesn't have any urban planners looking out for her. Her quiet, tree-lined neighborhood in Billings, Montana, looks exactly the same as it did twenty years ago. (The elms may be a bit taller, but that's it.) For Molly and her family, real change started at home. Her mom stopped picking up the family dinner at fast food restaurants, her dad stopped buying donuts on "special" Sundays (i.e., those that followed Saturdays), and Molly stopped eating the "three-napkin" pizza slices in the school cafeteria. (It took three napkins to soak up the grease.) A sack lunch with a turkey sandwich and an apple from home was a big improvement, she says.

Molly also stopped catching a ride to and from school. The six-block walk seemed tiring at first, but it soon became routine. She started spending more time climbing trees or jumping on her backyard trampoline and less time watching television. These changes in her life quickly led to a change in her size. In just half a year, she lost eight pounds while growing two inches taller. As her body slimmed down, her energy soared. She plans on spending much of her summer at the nearby outdoor public pool, and she hopes to have time to take up golf. When gym classes start up again in the fall, she won't mind waiting for the slow kids.

The kids running in the shadow of the Stratosphere don't have as many choices as Molly. They eat when they can, they play where they're allowed, and they go to school when they have to. And when they're in their favorite class, they do what Mr. Leon tells them. As the balls fly around on the well-baked grass, some of them just might develop a skill that doesn't involve a mouse or a joystick. A few might even discover that they enjoy running around and pushing their bodies. Far above them, another group of tourists takes a ride to the top of the tower. The view of Park-Edison elementary and beyond is impressive, even if it is a little hazy.

6

Obesity Goes Global

ON A RECENT AUTUMN AFTERNOON on New Zealand's North Island, a swarm of preteen school kids descended on a rocky beach. In their masks, snorkels, flippers, and light blue wet suits, they looked better equipped for a naval salvage mission than a gym class. After half-listening to their safety lecture and rushing through their equipment check, they crashed into the surf and started searching for sea life. While the other kids splashed and kicked and swam, one brown-haired girl was still stuck on shore, groaning and muttering. No matter how hard she pulled and tugged, she just couldn't get her wet suit over her belly. By the time two women came over to lend a hand, the girl's face was red with shame and exhaustion. Her two assistants—probably mothers of other students—grabbed fistfuls of neoprene and pulled upward as if lifting a large sack. The girl finally zipped up her zipper without a hint of triumph or relief. Wet suits are great for keeping out the cold, but they're no place for a fat kid to hide.

Throughout New Zealand, overweight children are easy to find. Native Maori kids in tiny villages, Samoan kids in metropolitan Auckland,

Caucasian kids on remote sheep farms: somehow, many children in this sports-mad country have gone from bantam weights to heavy weights. In 2003, the government announced the sobering results from a first-of-its-kind nationwide survey: more than thirty percent of kids were over the ideal weight for their height, and ten percent were extremely heavy.

Experts only recently started tracking the weights of New Zealand children, so nobody knows how much things have changed over the years. But for Professor John Birkbeck, director of the Institute of Food, Nutrition, and Human Health, the picture couldn't be clearer. He sees it in playgrounds and grocery stores, at sports stadiums and movie theaters. More and more kids—especially young Maoris and Pacific Islanders—are getting fatter and fatter, and the problem seems bound to get worse. "It's an exponential curve, and we're getting to the point where it starts to go straight up," he says. "It's urgent that we do something."

From the U.K. to Malawi

If it can happen in New Zealand—a country that invented bungee jumping, a country that put adventurer Sir Edmund Hillary on the $5 bill, a country that bills itself as a Mecca for sports and active living—it can happen just about anywhere. Indeed, the problem of childhood obesity has already spread across the globe. Between 1978 and 1995, the percentage of Egyptian preschoolers who are overweight nearly quadrupled. In the United Kingdom, the obesity rate for boys ages four to eleven nearly tripled between 1984 and 1994. From the mid–1980s to the mid–1990s, the percentage of overweight Scottish boys and girls roughly doubled. During the same period, the rate of obesity for Australian girls ages seven to fifteen grew nearly fivefold. Preschoolers in Ghana, teenagers in Brazil, city kids in China, American kids of all ages: they lead different lives, but they're all headed in the same direction.

Figure 6.1 Worldwide Obesity Rates for Children Ages Five to Seventeen Years

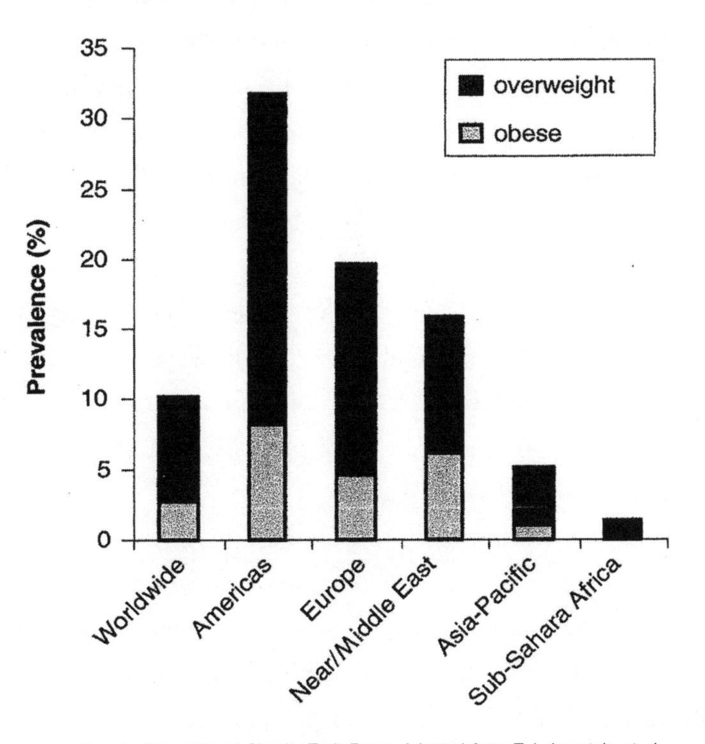

Source: International Obesity Task Force. Adapted from T. Lobenstein et al., "Obesity in Children and Young People: A Crisis in public Health." *Obesity Reviews* 5 (2004).

If obesity is a personal failing—as so many people seem to believe—the planet must be experiencing a huge, widespread, and simultaneous breakdown of self-control and good sense. But is it really possible that the same outbreak of sloth and gluttony has reached all the way from Philadelphia to Beijing? Have mothers and fathers in Malawi and Uzbekistan lost all good sense? Is Sri Lanka—a country where only one in one thousand preschoolers is overweight—the last bastion of self-discipline and effective parenting?

As childhood obesity spreads across the globe, the logic behind the personal weakness argument falls apart. Simply put, there's no way so many kids in so many diverse cultures could suddenly lose their willpower. Still, blaming the victims remains a popular pastime, even in the upper reaches of government. William Steiger, the George W. Bush administration's special assistant to the secretary for International Affairs, recently said that "personal responsibility" is the key to battling the obesity epidemic worldwide. It's a rather polite way of saying that heavy people have only themselves to blame.

"That's a very American approach," says Neville Rigby, director of policy and public affairs for the London-based International Obesity Task Force (IOTF). "It's impossible to see how you can make any changes at a population level just by telling people to make the right choices. We tell kids to make healthy choices, but we ask them to grow up as if they were living in a sweet shop."

Michael Lowe, a professor of psychology at Drexel University in Philadelphia and an expert in weight loss, has another term for the personal responsibility argument: "Bullshit." He elaborates: "The Bush administration is as least fifty years behind the times," he says. "It's the industry line: it's not our food, it's personal responsibility."

If an infectious disease reached all the way from Boston and London to Beijing and Tashkent in just a few years, public health experts everywhere would search frantically for the germ and all of its vectors—they wouldn't waste any time berating people for their weak immune systems. The hunt for the global causes of childhood obesity is just beginning, but a few things are already clear: there is no single culprit, there are no easy answers, and there is no point in blaming individual kids or their families. Something is going wrong with the global environment, and the children are suffering.

North and South, Rich and Poor

In this as in so many global trends, the United States—the country behind such innovations as chili dogs, cheese-stuffed pizza crusts, and sixty-four-ounce servings of soda—is leading the way. Still, the rest of the Western world isn't too far behind. According to the IOTF, roughly fifteen percent of ten-year-olds in the United Kingdom and thirty percent of ten-year-olds in Munich, Germany, are overweight. Even the food-conscious French are in trouble: a recent study of seven- to nine-year-olds in France found that more than twenty percent are overweight by CDC standards. Researchers noted grimly that the numbers put France on a par with the United States in the late 1980s, a precarious position to be sure.

A map of childhood obesity rates in Europe shows an interesting trend: the heavy kids seem to be collecting near the bottom. Overweight kids are roughly twice as common in southern countries such as Greece (thirty-one percent), Spain (thirty-four percent), and Italy (thirty-six percent) than in northern countries such as Sweden (eighteen percent), Denmark (fifteen percent), and Germany (sixteen percent).

The trend holds within individual European countries too. Children and adults in the southern part of Italy tend to be heavier than people in the north, says Italian obesity expert Francisco Branca, PhD. Nobody knows for sure why obesity rates change with latitude, but Branca sees a corresponding gradient in feelings about exercise. "It's a cultural attitude," he says. "In a recent survey of Europe, Italy and Portugal had the fewest people involved in organized sports." People in southern regions also spend less time working in the gardens or getting other types of physical activity, he says.

Obesity rates in Europe are also a reflection of wealth and poverty. Russia is one of the thinnest countries on the continent, and not just because it happens to lie in the north. According to the IOTF, the proportion of overweight children in Russia actually dropped from sixteen percent in 1992 to nine percent in 1998, a six-year span of severe economic hardship throughout the country. In a 2004 study that compared the weights of thirteen- and fifteen-year-olds in the United

States, Israel, and thirteen European countries, another impoverished country—Lithuania—had the fewest overweight kids.

Poor countries aren't completely immune to this new epidemic. In Uzbekistan, a nation lurching toward a market economy after decades of Soviet rule, fourteen percent of preschoolers are overweight and nearly that many are wasting away from malnutrition. Roughly seven percent of preschool kids are overweight in Malawi, a West African country with an annual per capita income of less than $200 (US).

Sweets in the City

The childhood obesity epidemic may have gone global, but large parts of the world remain untouched. Walk through a village in rural China, for instance, and you can go a long time without seeing a kid with a chubby face. You'll see kids working in fields and kids riding bicycles and kids pushing carts, but you won't see kids with huge bellies hanging out of their shirts. The scene changes dramatically if you visit the megalopolis of Beijing, a vast city of high-rise apartment buildings, endless traffic jams, and increasingly heavy kids.

A recent study by Harvard researchers documented an incredible upsurge in childhood obesity in Beijing and other Chinese cities during the last decade-and-a-half. Way back in the late-1980s, city kids tended to be slimmer than rural kids. Country kids are still skinny, but the city kids have gone on a mind-boggling weight-gaining binge. By 1997, obesity rates in urban preschoolers increased more than sixfold to nearly thirteen percent.

The explosion of childhood obesity in Chinese cities typifies a global trend, says Monika Blossner, a Geneva-based nutritionist with the World Health Organization (WHO). In countries across the world, families are moving from farms, villages, and small towns to big cities in search of jobs, better health care, education, and a new life. All too often, that new life includes a new opportunity to grow fat. You can see it in London or Los Angeles or Munich, especially in the poorest parts of town where buildings are crumbling and hamburger wrappers and plastic soda bottles drift up against the curbs. In New Delhi or Beijing or Nairobi, however, heavy children tend to live in the nicest neighborhoods. Only the relatively wealthy families in developing countries can afford the Western way of life, putting them at risk for Western-style obesity, Blossner says. "They see a lifestyle represented on television," she says. "If you're wealthy, you eat corn flakes and drink soda."

City living doesn't simply shape children's desires or ambitions—it also dramatically changes their options for food. Every large city in the world now offers plenty of quick calories on the go, often in the form of brand-name fast food. Big cities also offer people an incredible number of ways to satisfy their sweet tooth. Sodas and sweets are sold on many street corners, not to mention gas stations and laundromats and restaurants. City dwellers also have easy access to processed foods that are subtly or not-so-subtly sweetened with sugars, from cereals to white bread to potato chips.

There's strong evidence that the mass migration from the country to the city has sweetened the world's diet. In a 2003 study published in *Obesity Research*, world-renowned nutritionist Barry Popkin, PhD, estimated that the average person on the planet got seventy-four more calories each day from added sweeteners in 2000 than in 1962. Although the sweets came in many forms, sodas accounted for the lion's share of the extra calories.

Popkin, a professor of nutrition at the University of North Carolina at Chapel Hill, attributes some of that increase to good marketing—more people evidently decided that Coke really was "it." But his study uncovered an even more important trend: throughout the world, the boost in calories from sweeteners was tightly tied to growth of cities.

As more people left their farms and villages, they added sodas, processed foods, and sugary treats to their daily routine. Soda companies and sugar manufacturers call it a coincidence that obesity rates have climbed in lockstep with sugar consumption, but Popkin says the sweetening of the world's diet is a significant factor in the fattening of the world's children.

Nobody believes that sodas and sweets are the sole cause of the global epidemic. In every part of the world, children are a reflection of their environment. And in every part of the world, that environment is far more complex than the nutrition label on a can of soda. The causes of obesity are tied up in cultural beliefs, economics, and changing lifestyles. In many places, the conditions are ripe for disaster.

"Little Fatties"

The cultural forces driving childhood obesity are especially complex in China, a country where tradition is becoming intertwined with Western food and marketing. The Chinese have an affectionate term for overweight children—*xiao pangzi*, or "little fatties." Chubby little legs and cheeks have long been a sign of family wealth and good fortune, and Chinese parents and grandparents encourage their young ones to eat well and often. Ever since the government adopted its one-child law in 1979, the doting has grown even more intense. Most families have just one chance at raising a little fatty of their own, and they want to make the most of it.

Parents and grandparents still boast about hefty children, but attitudes are changing. In a country where one hundred million people have high blood pressure and twenty-six million have diabetes, round bellies just don't have the same cachet. Many parents have even sent their kids to one of many "fat camps" that have sprung up around Chinese cities. In the mid-1990s, Jung Shu Yi, a director of a fat camp, lamented the state of Chinese children in a broadcast on National

Public Radio. "One thing that's definite is that these children are fat because they're spoiled by their families or treated like little emperors," said Jung through an interpreter. "When they come here, they have a lot of bad qualities. For example, they are greedy and fussy about their food. They're lazy and bad tempered." It can only be hoped that someone else was in charge of camp morale.

Even if every parent and grandparent in China decided tomorrow to start raising "little skinnies," the problem of childhood obesity wouldn't go away. As Popkin and others have documented, urbanization and rising incomes have revolutionized the Chinese diet. Chinese families are eating fewer vegetables and less rice but more red meat, eggs, dairy products, sodas, and sugars, a trend that is boosting their intake of fats and calories and almost certainly expanding their waistlines. The revolution is partly homegrown: in the early 1990s, the government started supporting the local sugar and soda industries with loans and hefty subsidies. The subsidies made Chinese companies more competitive in the international market, but they also greatly increased the availability and lowered the price of sodas and sweets for Chinese families.

The most dramatic development in Chinese eating habits just might be the shift from home-cooked meals to Western-style fast food. You don't have to read Chinese to understand the cuisine options in Central Beijing: neon signs beckon customers to Starbucks, KFC, Pizza Hut, and (it probably goes without saying) McDonald's. McDonald's has forged an especially prominent niche in Chinese culture. It's not even seen as an American restaurant; to many kids in Beijing, "Chinese food" means two all-beef patties, special sauce, lettuce, cheese, pickles, and onions on a sesame seed bun. The restaurant offers reliable food and clean bathrooms, two things often lacking in traditional restaurants. "Consumers have turned their neighborhood restaurants into leisure centers for seniors and after-school clubs for students," states James L. Watson in *Foreign Affairs*.

McDonald's has been closing restaurants in other parts of the world, but one hundred new restaurants spring up in China every year, and the company aims to have 1,000 Chinese outlets in time for the summer Olympics in 2008. McDonald's still has plenty of competition, and not just from other American companies. Several China-owned chains—including McDuck's, Mordornal, and Mcdonald's—are tapping into the newfound love for hamburgers, fries, sodas, and shakes. (Evidently, China's trademark laws aren't evolving as rapidly as its tastes.)

Big Macs are just as fatty and full of calories in China as anywhere else, but they have earned one important distinction: they happen to be the cheapest on the planet. Economists use Big Macs to gauge the purchasing powers of different currencies—it's called the "Big Mac Index." The latest index shows that Big Macs in Beijing typically cost about $1.20 in U.S. dollars, less than half the price in the United States and cheaper than anywhere else in the world. (This fact gets economists riled up about the "undervaluation" of the Chinese yuan, but, for most people, it means nothing more than a good price on a burger.) That's cheap enough to put the burger well within reach of kids and teens from middle- and upper-class families. Youngsters file in for dinner or after-school meals, hang out at the well-scrubbed tables, and, slowly, calorie by calorie, a growing number of them take another step toward a weight problem.

From Starving to Stout

Across the continent, another country is raising its own crop of overweight children under very different circumstances. Uzbekistan, a dry, rugged country in Western Asia, has seen more misery than prosperity in recent decades. A few scattered oil wells and hefty military investments from the United States (a country with keen interests in Uzbekistan's neighbor, Afghanistan) have helped this ethnically diverse Muslim nation creep into the world of capitalism. Still, the main streets of Old Tashkent don't exactly have the bustle of a boomtown. Men ride donkeys to the market, and children in brightly colored wool caps and sweaters chase after beat-up bicycle-tire rims that are just barely round enough to roll. The few fast food restaurants in town are vastly outnumbered by sidewalk vendors tending steaming pots of traditional fare: rice mixed with chickpeas, carrots, and small bits of lamb.

Active kids surrounded by healthy food—at first glance, it's hard to see how one in seven youngsters in Uzbekistan ends up being overweight. Francisco Branca recently visited the country to put kids on scales, investigate their diets, and unravel this mystery. He concluded that an expanding economy has led to expanding waistlines. Oil wells and foreign investments have created a new upper class, and some money has trickled down to people on the street. While many kids are still starving—especially in areas of the country wracked by an ongoing drought—other children are starting to get the first regular meals of their lives. For rich kids—and surprisingly, for poorer kids—this newfound wealth often means excess pounds.

The story for wealthy families is straightforward and familiar. Flush with cash, they fill their lives with luxury items—including fatty foods—and revel in the sedentary life. "Upper-class families eat meat with practically every meal, including the fat," Branca says. "They also drive as much as possible. Cars are a major status symbol." It's no surprise that wealthy families often eat and drive themselves to obesity. Here and elsewhere, extra weight can become a status symbol in itself, Branca says. In parts of the world where many kids waste away from malnutrition, body fat is a badge of good health and good fortune.

Ironically, a growing number of Uzbeki children living in makeshift tents and dirt-floor hovels are sharing this "status symbol." The impoverished children of Uzbekistan put a different face on the epidemic of childhood obesity, and they represent the future of many developing countries. In poor regions throughout the world, millions of children and infants suffer from malnutrition. Many are malnourished at birth, and others quickly fall behind as toddlers or preschoolers. Monotonous meals of rice or bread may give them enough calories to get through the day, but the kids are missing out on vital micronutrients such as iron, zinc, and iodine. Their growth becomes stunted, and their metabolism slows down.

The consequences of malnutrition will reverberate for decades—for children lucky enough to live that long. If a malnourished child's diet doesn't improve, he or she will grow progressively weaker and become more and more vulnerable to infectious diseases. Malnutrition has another long-term effect on children: it "reprograms" their bodies to put on the most amount of fat with the least amount of calories. If starving children suddenly start getting meals that are rich in fat or calories, their stunted bodies will greedily put on weight. Malnourished kids who suddenly get a calorie windfall often develop the same type of "metabolic syndrome" that afflicts American teenagers with a love for fast food. As their waists expand, their blood pressure soars, and their bodies lose sensitivity to insulin, a one-two punch that puts them at risk for heart disease and diabetes.

In a report issued in February 2004, the United Nations' Food and Agriculture Organization (FAO) warned that vast numbers of malnourished children in developing countries could soon send the global epidemic of obesity into high gear. As high-fat and calorie-dense foods become more available, more starving kids will put on weight at incredible speeds. The resulting explosion of obesity, diabetes, and heart disease could cripple the healthcare budgets of many poor countries. Nobody is suggesting that starving kids should be left to starve. Quite the opposite: According to the FAO, the specter of obesity gives the entire world a new reason to step up efforts to prevent malnutrition in pregnant women and young children. The title of the FAO's report speaks volumes: "Fighting Hunger Today Could Help Prevent Obesity Tomorrow."

Behind the Curve, Moving up Fast

Far to the north, another country is seeing an unexpected surge in childhood obesity. Sweden is often held up as an example of a country with every possible advantage in the fight against obesity. Swedish kids don't watch nearly as much TV as their American peers, and when they do, they aren't bombarded by ads for choco-cocoa-bomb cereals or fantastic-frooty-tooty-sugarlicious drinks. (Swedish television prohibits ads aimed at children.) Swedish kids also walk or bike to school—a lost art in many other places—and they eat healthy lunches when they get there. They even have regular, vigorous PE classes. So when kids in Sweden start getting heavier, the world should take note.

And, without a doubt, kids in Sweden are putting on weight at a brisk clip. A study conducted as part of the Stockholm Childhood Obesity Prevention Project (STOPP) found that the percentage of overweight seven-year-olds in Stockholm increased from eight percent in 1989 to approximately twenty percent in 2004.

Although some people continue to laud Sweden as a model for healthy living, Claude Marcus, a pediatrician and director of STOPP, has treated too many 200-pound kids to believe in this romanticized ideal. One boy in particular sticks in his mind. He was a typical Swedish child with typical Swedish habits. He played soccer. He loved hamburgers. His parents were fit and active. And when he stepped into Dr. Marcus's clinic at the age of eleven, he was at least one hundred pounds overweight. Dr. Marcus quickly realized that this young boy wasn't doing or eating anything wildly out of the ordinary. "Most overweight kids [in Sweden] eat essentially the same foods in the same amounts as lean kids," he says. The kids with a genetic predisposition toward storing fat—a common trait in Scandinavia—are the ones most likely to show up in his office, he says.

In the last couple of decades, Marcus says, young people in Sweden have radically changed their approach to eating. Twenty years ago, they had few options for fast food: it was basically hot dogs, hot dogs, or hot dogs. Needless to say, hot dogs lost their luster after a while. As Marcus puts it with Scandinavian understatement, "When you only have one type of fast food available, interest is going to be limited." Today, kids can get all kinds of high-fat, high-sodium, high-sugar meals on the go. They also get plenty of sugary treats at school, a place where ninety-five percent of Swedish kids are sequestered from morning until early evening, including classes and after-school care. Lunches are still healthy, but many kids wash down their lean meat and vegetables with soda from a vending machine. The calories really pile on when classes are over, Marcus says. Cookies, candy, cake, chips, and soda are all standard fare during after-school care.

Swedish kids have also discovered new types of leisure-time fun. They may not watch much TV, but they know their way around a joystick or a mouse. "Sweden is highly computerized, and just about every child has access to computer and video games," Marcus says. You still see kids playing soccer in the parks and racing bicycles on the streets, but a lot of the real action these days takes place on a screen.

New Zealand is another country where the sudden abundance of overweight children defies easy explanation. Change comes slowly here, and on the surface at least, kids today don't seem much different from kids twenty years ago. But young Kiwis (as New Zealanders call themselves) do pay attention to what's going on overseas, especially in a certain country to the distant northeast. "Kids are strongly influenced by American culture," says Lorna Gillespie, a high school physical education instructor in the city of Christchurch. "We're sort of insecure. If something is happening overseas, we think we should be doing it here."

Like other kids across the globe, young New Zealanders have embraced American fast food, but it's hard to say that their diet has taken a drastic turn for the worse. Twenty years ago, every New Zealand town had at least one "takeaway" shop that sold hamburgers, fish and chips, and an assortment of other deep-fried treats such as crab sticks and paua (abalone) fritters, green, rubbery discs that could stand as a dictionary definition for "acquired taste." The enticing smell of crisp batter and salty grease hung in the air of most street corners, and few kids—or adults—could resist it for long.

The shops and the smells are still there, and the menus haven't changed. (In many places, menus *literally* haven't changed: they're painted on walls and signs so old that the letters in "crab sticks" and "fish and chips" have started to peel away.) McDonald's and KFC have recently stolen many customers from the takeaway shops—especially in the cities—but anyone who switches from a deep-fried paua fritter to a Big Mac isn't making a huge nutritional sacrifice.

Even though the New Zealand diet hasn't gone through a major upheaval on the scale of China or even Sweden, young Kiwis still have new ways to get empty calories. "Kids ate a lot of fish and chips and drank a lot of Coke in the past, but now there are a whole multitude of other temptations," says Celia Murphy, head of a Wellington-based organization called Fight the Obesity Epidemic. "There are lots of sweets, energy drinks, and convenience foods. Young people today also have more money and independence, so they're making their own food choices. That's why obesity rates are getting worse as the kids gets older."

And even though Kiwi kids are just as sports-mad as ever—twelve-year-old boys and girls still get up in the middle of the night to watch the beloved All-Blacks play a live rugby match in England—these days not all kids are making the cut. "There's been a trend toward elitism," says Professor Birkbeck. "Children who are good at sport get encouraged, but those who aren't so good are actually discouraged." And although older kids still have vigorous PE classes, twenty percent of elementary school kids have no PE at all. In most New Zealand elementary schools, one teacher covers every subject, from art to math to music to PE. These teachers don't always have the training or even the stamina to put an emphasis on physical activity, Gillespie says. Their average age is forty-seven and climbing, and they're unlikely to rediscover the importance of exercise without a lot of outside help, she says.

The World Wakes Up

From New Zealand to Uzbekistan, governments, health experts, and families are waking up to the growing problem of overweight children and vowing to do something about it. So far, there is more talk than action, more good intentions than practical solutions. But the framework for change is already in place, and many people see room for optimism. Kids can eat smart, live well, and lose weight, but they can't do it on their own.

In Sweden, Dr. Marcus believes a few simple measures could put the brakes on the epidemic. The main battlefront will be the schools, where kids get about thirty-five percent of their total calories. He has already put his ideas to the test. In an ongoing study, he and his research team tweaked the aftercare diets of six- to ten-year-olds at five different schools. They cut back on sweets and soft drinks and offered more fruits and vegetables. After one-and-a-half years, the obesity rates at these schools didn't budge, whereas rates in similar schools steadily climbed. If all schools took such steps, he says, many fewer kids would ever have to visit his clinic.

The fight against childhood obesity is heating up on several fronts in New Zealand. In a rare show of solidarity, both the minister of health and the Food and Grocery Council—the face of the food industry—have pledged to do their part. In 2003, Health Minister Annette King announced a national strategy to encourage children and adults to exercise more and eat a more nutritious diet. Among other things, the government plans to fund walking programs and actively promote fruit and vegetables.

The Food and Grocery Council is taking a much more unexpected—and potentially more effective—approach. The council reached a turning point at a recent private meeting attended by Birkbeck. "They all sort of acknowledged that there was a problem and that maybe the food industry was partly responsible," Birkbeck says. "I told them the industry should adopt a code of practice. They should voluntarily examine all products for composition and serving size, and ask: what could we do differently?" The council quickly agreed to the suggestion. It remains to be seen, however, if the pledge will lead to less fat in the fish and chips, less sugar in boxed cereals, or any other positive change. Cooperation between industry and government—if it turns out to be more than an illusion—would surely set New Zealand apart from much of the rest of the world.

In other parts of the world, bold measures to combat childhood obesity tend to end up in the dustbin. In 2003, Debra Shipley, an English labour MP, introduced a bill to ban TV ads for high-fat or high-sugar foods during children's programs. Shipley noted that an hour of children's programming in Great Britain contains up to eleven ads for unhealthy foods. Many of the ads feature favorite television characters such as the Teletubbies or celebrities such as David Beckham, and they're all carefully designed to get kids to lobby hard for the treats in their next visit to the grocery store. The bill was supported by one hundred other MPs and many health organizations, but it fell by the wayside in March 2004 after stirring up strong opposition from the food industry. The government's official stance was that the main cause of childhood obesity was a lack of exercise—not overeating. As Culture Secretary Tessa Jowell put it in *The Express*, "We are getting fatter because we are less active. Of course advertising has an impact, but what we have to judge is whether a ban would be appropriate."

The United States Versus WHO

While countries here and there are tackling their small piece of the epidemic—or at least making noise about it—the World Health Organization is thinking big. For one thing, the WHO is taking the lead in addressing nutritional problems in the developing world. While industrial countries struggle with the dangers of Playstations and sugarcoated cereals, poor countries must confront the rampant malnutrition that stunts kids' growth and wreaks havoc with their metabolism, Blossner says.

There's little hope of stabilizing the global food supply any time soon, but families in developing countries can learn to make better use of what they already have, says Blossner. Workers with the WHO have been traveling from village to village in Asia and Africa to encourage mothers to breastfeed their children and to seek out a wider variety of foods that offer the right combination of nutrients. "Studies show that children are less likely to suffer from malnutrition if their mothers have been exposed to nutrition education," Blossner says.

The WHO is also gearing up to take a worldwide approach to the worldwide problem. In late 2003, the WHO released the draft version of a sweeping "Global Strategy on Diet, Nutrition, and Health," designed to prevent the worldwide spread of obesity, especially among children. The recommendations wouldn't be legally binding in any country; instead, they would provide a framework for nations concerned about obesity. Among other things, the WHO recommended curbing advertising aimed at children, imposing taxes on junk foods and soft drinks, using subsidies to lower the prices of fruits and vegetables, and encouraging the fast food industry to cut back on harmful ingredients such as trans fats. The document also referred to a previous WHO report that called for people to get no more than ten percent of their calories from sugar.

Health experts lauded the proposal, but it still ran into a monumental roadblock, the kind of diversion that only a superpower could build. In early 2004, the nation that leads the world in childhood obesity decided to lead the fight against the WHO's global strategy.

While U.S. Secretary of Health Tommy Thompson said in public that he was "very much in favor" of the global strategy, his department worked behind the scenes to undermine the proposal. In January 2004, his assistant William Steiger spelled out a long list of objections and criticisms in a thirty-page letter addressed to the director general of the World Health Organization. For one thing, many of the WHO's recommendations were not supported by science, at least in the eyes of the United States. The letter is full of comments such as: "No data have yet clearly demonstrated that the advertising on children's television causes obesity," and "There is only one study of the relationship of soft drinks and juice to obesity in children . . . the data do not provide sufficient support to be labeled probable."

Just weeks after the letter was delivered, both the American Psychological Association and the Kaiser Family Foundation released major reports linking advertising to childhood obesity. As for that "one study" on the association between soda and obesity in children: it was the study conducted by researchers at Harvard, the study that found each daily serving of a sugary drink raised the risk of obesity by sixty percent.

The U.S. government had another major gripe with the WHO report. As William Steiger said in interviews, the world should shift its focus from government-imposed solutions to "personal responsibility." In February 2004, the Department of Heath and Human Services suggested multiple revisions to the strategy, including inserting the words "personal" or "individual" nine times.

The attack riled public health experts across the country and across the world. The general consensus was that the United States seemed more interested in appeasing the food industry than in battling obesity. "The U.S. incorporated industry language into its critique practically word for word," says New York University nutritionist Marion Nestle. Neville Rigby of the IOTF agrees that the food industry—most notably the sugar industry—drove the Bush administration's rebuke of the global strategy. "The sugar industry may not have directly contributed to the critique, but they provided the motivation," he says. Michael Lowe of Drexel University sees a copout: "The government doesn't want to offend food companies, but they could do a lot better than throwing their hands up and saying it's personal responsibility."

Bill Pierce, a spokesperson for the Department of Health and Human Services, denies that Big Sugar or anyone else in the food industry had any input in the U.S. position. "Nobody can show me any evidence of that," he says. "They [the critics] have a political agenda that they're trying to drive. We didn't collaborate with industry in our response. Our response was put together by scientists. If countries want to pursue any policy, that's fine. But the WHO is a scientific body. The claim that advertising causes obesity is not an established claim. It may seem true, and some people may want it to be true, but it hasn't been scientifically established. One study does not make science, especially when you are making a policy decision for the world."

In the end, the World Health Organization refused to be rattled. The agency unveiled the final version of its strategy at the World Health Assembly in May 2004 with few concessions to the Americans. The final version still called on food manufacturers to improve the quality of their foods and to limit marketing to children. The sugar industry did score one victory, however. The final global strategy made almost no mention of the WHO report that called for people to get fewer than ten percent of their calories from sugar. Instead, the strategy merely recommended "limiting sugar," which is a bit like replacing a "Speed Limit 55 mph" sign with one that just says "Speed Limit."

The United States approved the final version of the strategy, and Thompson called it "good news for those countries around the world who are facing the issue of overweight and obesity." Of course, there's a huge gulf between complimenting a proposal and actually putting any of it into action. Nestle, for one, doesn't think that the WHO's strategy will have any effect on U.S. policy. Still, she says, the exercise may not have been entirely fruitless. "I think the government was probably surprised at the outpouring of protest about its shameful actions in protecting the health of sugar producers over the health of children," she says. "Perhaps government officials learned something. One can only hope."

7

Parents: What Helps, What Hurts

THE GRAND TOUR OF BUTLER, ALABAMA, is a six-stoplight trip. You'll pass a giant yellow smiley face painted on a sky-blue water tower, and you'll see about a dozen abandoned buildings, former banks and hardware stores and hair salons that shut down years ago. You won't see any movie theaters or shopping malls. You also won't see many kids riding bikes or playing catch—in fact, you probably won't see many kids at all unless you happen to drive by one of the town's two schools at recess. If you do glimpse a group of kids milling about, you'll realize that this small town just east of the Mississippi border has a big problem. It may be remote by most measures, but Butler is smack in the middle of the childhood obesity epidemic.

In the halls of the Patrician Academy, Butler's private school, sixteen-year-old Amber Kearley hardly stands out. She has short light-brown hair, brown eyes, and the same school uniform as everyone else. She's quieter and more polite than most tenth graders. At 5'3" tall and 160 pounds, she's a little on the heavy side, but that doesn't set her apart in a school full of chubby kids. What does set her apart is that she's doing something about her weight. Since seventh grade, she has dropped fifty pounds and six dress sizes while growing two inches taller, and she's still heading in the right direction. She's a forward on the school basketball team, and she has more stamina and speed than ever before. She's no star—shooting free throws, she says, is what she does best—but she's had her share of victories.

It isn't easy losing weight in the Deep South. The towns are blazing hot during summer and into the fall, and there aren't many bike lanes or walking paths. The closest thing to a walking path in Butler is a dirt road surrounding an abandoned clothes factory. And then there's the food. Although home-cooked meals still feature traditional sides of vegetables such as collard greens and succotash, just about anything else that can fit on a plate is in danger of getting deep fried. Recent innovations include the deep-fried turkey, deep-fried dill pickles, and deep-fried cheesecake. Southern culture is especially hard on growing children, says Louisiana State University's Melinda Sothern. "In the South, food is love," she says. "The more you feed your children, the better mom you are."

Butler is certainly no enclave of good nutrition. On a Sunday morning, the local Hardee's is the sole choice for breakfast. Biscuits dominate the menu: biscuits and gravy, ham and cheese biscuits, and the sausage and egg biscuit. For those watching their carbs, there's a biscuit-free option: an enormous pile of eggs, cheese, ham, bacon, and sausage. For lunch, there are sandwiches from Subway, fried chicken from Church's, and deep-fried catfish sold from a trailer sporting a mural of firefighters at Ground Zero. Bargain hunters may want to try the fried chicken at the local Piggly Wiggly or the Hot Stuff pizza at the Exxon station: buy a large pizza and get a free two-liter bottle of Coke or Pepsi.

If a city planner had set out to design an environment that would make everyone fat, the town of Butler, Alabama, would be close to perfect. How, then, did Amber Kearley defy the odds? You can find the answer if you happen to see Amber hiking around that abandoned clothes factory. She'll be walking with her mother, the exercise partner who's the key to her success.

Amber says she couldn't have shed a single pound without her mother's help. "She told me I could do it," Amber says. Susan Ryals, a single mother to Amber and her thirteen-year-old sister Kindall, works in the shipping department of the Georgia Pacific paper mill. The family lives in a modest prefabricated house shaded by tall pine trees. Susan works plenty of late nights, but when her kids need her, she's there. In fall 2001, she started driving Amber once a week to nutrition classes in Meridian, Mississippi, a solid forty-five minutes down the road. Amber was the only girl in a class full of women, but she proved to be a top pupil. With her mother by her side at class and at home, she lost twenty pounds in just a few months, and her weight has steadily declined ever since.

Working together, this mother and daughter have done something truly amazing. In the fattest region of the world's fattest country, they have beaten the odds. Susan's own weight-loss story is almost as dramatic as Amber's. Susan lost sixty pounds eight years ago through diet and exercise, and today she has the slender good looks of someone who never had to worry about weight. (Her face is glowing red as she heals from laser reconstruction surgery; she was tired of looking at those loose flaps of skin left over from her "fat" days.) Amber's still a little on the heavy side, but she's already thinner than many of the other kids at her school. A healthy, confident, athletic-looking teenager, she bears little resemblance to the chubby-cheeked sixth grader staring out from the pictures hanging in the family living room.

Amber and Susan may be exceptional, but they're also emblematic of a larger trend. Whenever you find an overweight youngster who has managed to slim down, you almost always find a supportive parent (or two) who is dedicated to creating a healthy home environment. Susan

Ryals and other parents across the country are starting to fight back against childhood obesity—and many are winning. They aren't putting their kids on radical diets or sending them on forced marches. They're attacking the problem with slow, steady, incremental changes, and their kids are slimming down.

Time to Act

Amber was slender as a young girl, and if her life had just gone a little more smoothly, she might have stayed thin. But in 1998, both of Susan's parents—Amber's much beloved grandparents—were killed in a car crash in Hattiesburg, Mississippi. The next year, Susan divorced Amber's father. Like many other kids facing major upheavals, Amber found comfort in a steady stream of snacks and junk food. "I would eat a bunch of chips and then move on to something else," she says. "It was like I couldn't decide what I wanted." The junk food helped her get through the day, but it also transformed her body. Throughout fifth and sixth grades, Amber put on weight at an alarming rate.

Even as the pounds added up, Amber never seemed concerned. She never said she wanted to fit into smaller clothes or look like her "normal" friends at school or just have a thinner body. In retrospect, her acceptance of her body was an act. "She's really good at holding things in," Susan says. The truth came out when Amber and her younger sister started seeing a child psychologist to help them deal with the lingering grief over the loss of their grandparents and the divorce. In just two sessions, the psychologist uncovered Amber's hidden anguish. Yes, she was extremely uncomfortable with her body, and yes, she wanted to get help.

Susan had worried about Amber's weight long before this revelation. Susan used to work as a respiratory therapist in a hospital, where she saw the consequences of obesity firsthand. "I saw a bunch of kids who were overweight and had health problems like diabetes or heart trouble," she says. "I didn't want that for Amber." Still, she felt she couldn't help Amber if she didn't want help. When she realized her daughter wanted to lose weight, she knew it was time to act.

Karen Rodgers of Redwood City, California, has been worried about her son Jake's weight pretty much since the day of his birth. He was born a month early but still tipped the scales at seven pounds, four ounces. As he grew older, he had a seemingly unstoppable tendency to pack on the pounds. Serving him healthy food didn't help, Karen says. He'd gag every time he ate a vegetable. He wasn't crazy about fruit either. Cheeseburgers, however, became number-one on his food list. The ten-year-old is 5'3" and weighs 180 pounds. The good news is that, like Amber, he has a mother who is committed to helping him slim down. Karen—a management analyst for the county coroner's office, a part-time police dispatcher, and a single mom—is just starting this journey, but she's taking all the right steps.

An easygoing fifth grader with a chubby, round face and shaggy, light brown hair, Jake has had the good fortune to escape cruel teasing by other kids. He knows he's heavy, but he doesn't see a problem with it. On a warm summer day, he wears shorts and a loose T-shirt and sees nothing about himself he doesn't like. But exercise doesn't come easily to him. At ten, he lacks the muscle coordination to ride a two-wheeler. And his mother worries that as Jake gets older, he'll face an onslaught of taunts about his weight that will undermine his self-esteem. Worse, she fears, if he remains as overweight as he is now, he'll be a strong candidate for diabetes and heart disease.

"I look at him, and I feel bad," Karen says. "Why did I let him get like that?"

Fatness and the Family

Not long ago, doctors expected kids to take full responsibility for their weight. They'd lecture young patients on the dangers of extra pounds and urge them to get more exercise. Some committed doctors would even send kids to see a registered dietitian, where they would learn about counting calories and the food pyramid and the dangers of fatty foods. All the while, their parents were somewhere else: in the waiting room, perhaps, or at home, or even at a nearby KFC. The kids may have become experts at nutrition, but, not surprisingly, they weren't always champions at weight loss.

But doctors' attitudes changed after the publication of a landmark study in the *Journal of the American Medical Association* in 1990. Researchers from the University of Pittsburgh put overweight children and their parents through three educational programs, but only one involved both parents and children. Ten years later, in the group that took a family approach to the problem, the proportion of children who were overweight had dropped approximately eight percent. In contrast, the rates of obesity had climbed significantly in the other groups.

The study energized obesity researchers everywhere. Doctors and dietitians who focus all their energy on overweight children realized they may be missing the real target: parents. Parents have more control over a child's weight than anyone else, and that includes doctors, nutritionists, friends at school, and even the child herself. No matter how deeply kids yearn for a slimmer body, it's hard for them to get anywhere without the support of the people who buy the groceries, make the dinners, and set the rules.

Of course, not even a dedicated mother like Susan Ryals or Karen Rodgers can make a child instantly thin, but parents everywhere can make a difference. In fact, they are often the only ones who can. Even in the face of all the unhealthy influences in the world, parents have an astonishing power to shape their kids' eating habits, says Moria Golan, PhD, a senior teacher in the School of Nutritional Sciences at the Hebrew University of Jerusalem. Fast food companies could start building new restaurants on every corner, spend billions more on advertisements, and offer two or three toys with every meal, but they can't tell parents what to put in the refrigerator or on the dinner table. Childhood obesity is an environmental problem, Golan says, and parents are still a major part of any child's environment.

Golan has long believed that parents hold the key to childhood weight loss. In 2004, she published a study that underscored her message. The study involved sixty overweight children ages seven through twelve. Half of the kids attended thirty intensive sessions where they learned about diet and exercise. The second half stayed home while their parents went to fourteen classes where they learned how to encourage children to eat well and stay active. Both groups of kids slimmed down after one year, but the kids who stayed home while their parents went to class lost significantly more weight. After three years, things really got interesting: the kids who had attended the thirty educational sessions were more overweight than they were before, while the kids who stayed home had moved even closer to their ideal weight. The message is clear: when parents are committed to creating a healthy home environment, kids can lose weight and keep it off.

The kids who attended the thirty educational sessions obviously knew plenty about exercise and nutrition—they could probably write a book about it—and they were highly motivated to lose weight. So why did they fall so far short of their goals? During sit-down talks with all of the kids, Golan heard the same story over and over again. "They were dieting all of the time," she says, "but that's not the solution." Almost without exception, the dieting went hand-in-hand with binges. The kids could deny themselves fatty foods or sweet treats for a while, but eventually they let down their guard and ate more than they had before. A couple of girls in this group regularly forced themselves to vomit after their eating sprees, an eating disorder known as bulimia or "bingeing and purging."

Josh's story: "We had to make small changes"

In some ways, getting diagnosed with prediabetes and high blood pressure was the best thing that ever happened to sixteen-year-old Josh Graham. Before that day in February 2003, he had no clue that his weight was seriously threatening his health. The teenager from Clinton, Connecticut, carried 420 pounds on his six-foot frame, and he probably would have just kept getting bigger without a serious wakeup call.

He had passed the 200-pound mark in grade school. He topped 300 pounds in middle school. By the time he was a sophomore in high school, he was heavy enough to bury the needle on the bathroom scale.

Josh's issues with his weight started when he was very young. Even as a grade schooler, he couldn't ride the school bus without the other kids teasing him mercilessly. The ridicule subsided as he got older, but his self-consciousness only grew. By the time he reached high school, he refused to go into public places by himself, not even the nearby convenience store. If people were going to stare at him—and they would—he wanted his mom or dad or older brother by his side.

The news that Josh was at risk for diabetes shook his entire family. Although he is of Polish descent—a group not at especially high risk for diabetes—three of his uncles and a grandmother were already living with the disease. Everything appeared hopeless: he couldn't walk more than a few hundred feet without getting winded, so how could he ever start exercising? He had a seemingly limitless appetite for hamburgers, french fries, soda, and potato chips, so how could he possibly switch to produce and whole grains? Besides, his dad, Bob, weighed more than 400 pounds and his mom, Tracy, weighed nearly 300 pounds. Biology was clearly working against him. How was he going to fight it?

Unlike many kids on their way to diabetes, Josh grew up in a middle-class household. His family could afford medical care, and they made the most of it. Josh's doctor referred him to Bright Bodies, a 12-week pediatric weight-loss program at Yale–New Haven Hospital in New Haven, Connecticut. Like other successful programs, Bright Bodies is a family affair, offering nutrition classes for parents and kids, along with exercise sessions and parenting classes.

At first, Tracy doubted that any program could help Josh. "I really didn't think he could lose a significant amount of weight," she says. The program has an answer to such skepticism. Tracy was advised to tweak the family's diet in just one way for two weeks before starting the program. So the first thing she did was replace the fruit juice everyone guzzled throughout the day with water. After that one minor adjustment, Josh lost four pounds in two weeks. "That's when I realized that we were just going to have to make small changes," Tracy says.

As the program got into full swing, Tracy was inspired to clear out the fridge and the kitchen cupboards. The local food bank got a windfall of cookies, crackers, chips, and sugary cereals, and the Graham family got a start on a healthier life.

➔

Josh still eats as much as he wants—Tracy says she didn't want to limit his food—but now he fills up on such things as baked chicken with vegetables and snacks on low-carb peanut butter sandwiches. He has also started to steer clear of the cafeteria food, something that admittedly didn't take that much willpower. He says the food at his school is really bad, but he's practically the only student who brings a sack lunch.

The whole family has also made a new commitment to exercise. To everyone's surprise, Josh has turned out to be a real athlete. The teenager starting logging slow, steady miles on a treadmill. "At first I could only go ten minutes or so before getting out of breath," he says. Soon he was able to get up to twenty and then thirty minutes. Walking in place helped him get into shape, but it didn't exactly get his adrenaline pumping. He quickly took up a new hobby that few 400-pounders would ever consider—boxing. Working with a personal trainer, he pounded a punching bag, learned footwork, and sparred a few rounds in the ring.

Josh is still a heavyweight, but he feels like a whole new person. The new Josh is 100 pounds lighter. (For the record, the new Tracy is 110 pounds lighter, and the new Bob is fifty pounds lighter. Between the three of them, they lost the equivalent of an NFL linebacker in just over a year.) His ultimate goal is to get down to about 250 pounds, and his mom is encouraging him every step of the way. Some of her methods are a bit unorthodox, however. After reaching a certain weight, Josh was allowed to get a new skull-and-crossbones tattoo. If he loses another fifty, he'll be able to modify an old car into a racer.

But weight loss has already changed the young man's life. He's fitter, and he no longer has either high blood pressure or high blood sugar. He'll never score a more important knockout. •

None of the kids in the group that stayed home developed bulimia or any other eating disorders. In fact, they rarely felt the need to binge at all because they weren't denying themselves anything. They were just eating what was around the house and going with the flow—a flow that happened to carry them to a slimmer, healthier life.

Golan believes all childhood weight-loss programs could focus entirely on the parents and exclude the children—after all, she says, it's the environment that needs to change, not the kids—but few other experts are willing to go out on that particular limb. Most encourage a family approach where parents and children work together. Melinda

Sothern's Committed to Kids program, developed at Louisiana State University, requires family participation. So does Shapedown, a national program for overweight children and their families, as well as Yale University's Bright Bodies program.

Sothern and her colleagues at Committed to Kids enlist the support of parents and siblings from the very beginning. If the parents happen to be overweight themselves—and many are—the staff strongly encourages them to go through every step of the program alongside their kids. A nutritionist helps them rethink their diet, an exercise physiologist helps them get moving, and a counselor helps resolve the emotional issues that are so often wrapped up in weight. Even if these parents don't lose much weight on their own, their dedication sets an example for their kids, Sothern says. Time and time again, Sothern has seen kids lose impressive amounts of weight with their parents by their side.

Before parents of an overweight child can put their power to good use, however, they have to overcome a nearly universal obstacle: guilt. Sothern senses that emotion just about every time a mom or dad enters her office. "The first thing I tell parents is 'it's not your fault,'" Sothern says. "It's your responsibility to make things better, but it's not your fault."

The message can take a while to sink in, she says. Some of her parents are so wracked with guilt that they're afraid to seek help: what if the doctor calls the police, and they lose their child? In the mind of a worried parent, letting a kid get fat may feel like a felony, and in rare cases the law might actually agree. In one recent case, a single mother was convicted of misdemeanor child neglect after her 672-pound, thirteen-year-old daughter died of heart failure.

Parents who blame themselves need to be reminded that the cards are stacked against them, Sothern says. "Nothing in society is going to help parents who are trying to help kids lose weight. Everything is designed to make kids fatter."

In many cases, parents know only too well what their children are going through, says Martha Lee Palotta, a registered dietitian who runs a Shapedown program in the New Orleans suburb of Metairie, Louisiana. They know what it's like to be the fat kid on the playground, the kid who gets teased in the cafeteria, the kid who gets picked last in gym class. As they see their own children get fatter and fatter, they feel helpless and ashamed. When parents come into her office, it doesn't take long for those feelings to come out. "I keep a box of Kleenex right by my desk," she says. "If just two people cry in a day, I think it's a good day."

At the other end of the spectrum, a few parents come to Palotta with a completely clear conscience. They see themselves as blameless victims of someone else's weakness. A twig-thin mother wanted Palotta to "do something" about her very slightly overweight daughter. "I promised myself I would never have a fat child," the mother said during a private session. Palotta says that while the child didn't hear the comment, she probably gets the message loud and clear every day. "You can't hide that kind of attitude," she says. "It hurts the child, and it compounds the problem. Emotional issues like that often lead to overeating."

Many parents also worry about when they should, well, worry. Should they be concerned about the baby with fat legs? What about the five-year-old with a pot belly? Sothern says parents shouldn't be too concerned about a chubby infant or toddler, unless the child seems lethargic or has an insatiable appetite. (Insatiable appetite in a toddler may be a sign of a serious medical disorder such as Prader-Willi syndrome.) An overweight preschooler shouldn't cause much alarm either, assuming neither parent has a weight problem. But if there's a family history of obesity, it's a good idea to ask a pediatrician for advice. If a child is still overweight at age five, parents shouldn't expect him to "grow out of it," she says. At this point, it's definitely time to take action.

"It's Not Easy Stuff"

On a July evening, in the middle of a busy work week in Palo Alto, California, seven overweight kids and their parents line up to get weighed before a Shapedown meeting begins. As they squint hopefully at the scale, there are grunts of disappointment from some of the kids but also words of encouragement from their parents. The children range in age from eight to fourteen and from slightly overweight to quite heavy. Nearly half the parents are overweight as well.

The kids and their parents—including Karen and Jake Rodgers—are each working through separate workbooks in which, each week, they record what they eat, how much exercise they do, what their goals are, and how they're planning to meet them. Shapedown, a ten-week weight-loss program for children, isn't based on dieting; it's more of a healthy living program that encourages families to make lifelong changes that will affect their weight and health.

When the kids left the room for a while, the parents got a chance to vent. One mother talked about how hard it is to limit her chubby eight-year-old's consumption of cookies, candy, and chips, especially if the family's at a party or picnic where the parents can't control every bit of food available to their youngster. "I hate being the food police," she says.

Facilitator Anne Chasson, a nutritionist with a degree in counseling, has heard much of this before, and she encourages the parents to stay positive, to keep a sense of humor—and to be realistic. All children, she says, need to learn to make choices on their own and to care enough to make good choices. She tells them to keep modeling healthy behavior. "It makes a difference. Let them see you exercise. Lots of times we exercise when the kids aren't around. Eat healthy yourself. . . . That's the best we can do—to keep our home healthy." Chasson, who's trim and petite, said she's taught her two teenage boys—one of whom is fifty pounds overweight—to read labels every time they are out shopping. "Even if your kids don't buy into all of it, they may buy into part of it. It's not easy stuff, getting our kids to care."

Relearning How to Eat

Parents often feel besieged in a world of fast food outlets, cereal ads, and omnipresent vending machines. For many, the natural reaction is to launch an all-out counterattack against a fattening culture. They try to take total control of their family's meals: they choose the foods, they dish out the portions, and they decide when everyone else is really "full." They also strictly forbid any of the sodas, chips, and desserts that cause so much trouble in so many other families. And, far too often, they end up completely befuddled when their children grow fatter and fatter.

Well-meaning parents who micromanage their children's diets and strictly forbid certain types of food may actually be doing more harm than good, says Jennifer Fisher, PhD, of Baylor College of Medicine. Along with her colleague Leann Birch, Fisher has conducted several groundbreaking studies on the factors that shape children's eating habits. One recent study found that kids ate twenty-five percent less when allowed to choose their own portion sizes and consumed significantly more food when served supersized portions of their normal entrées. Another found that young children are more likely to snack when they aren't hungry if parents take strict control over meals.

Overall, it appears that kids may be more likely to gain weight if their parents take total control of mealtime and forbid certain types of food, Fisher says. "It's intuitive to think that when you don't want kids to eat something, you don't give it to them," she says. "But when kids have lots of restrictions, they don't grow up knowing how to be moderate eaters." When they run across forbidden foods outside the house—as they inevitably will—they tend to make up for lost opportunities, and then some.

Simply put, kids crave forbidden foods. For proof, you don't have to look any further than the "Goldfish Experiment," a scientific endeavor demonstrating, among other things, that studying preschoolers is a tough way to make a living. Fisher and Birch gave a box of Goldfish crackers and a box of wheat crackers to a group of preschoolers. Not surprisingly, the kids preferred the Goldfish crackers, and each happily snapped up handfuls of the treats. On another day, the researchers came back with the same goods, but this time the Goldfish crackers were locked away in a tub. The kids could have all of the wheat crackers that they wanted, but the Goldfish crackers were OFF LIMITS. A near riot ensued.

A few of the children started banging the their fists on the table and chanting "Goldfish, Goldfish, Goldfish." As the unrest mounted, a three-year-old girl stood up on a bench and shouted, "I'm out of here! Who's with me?" Fisher watched in disbelief as several kids bolted for the door. ("I thought, 'Oh no, there goes my data.'") After chasing down her data points and herding them back into the room, Fisher finally opened the lid on the Goldfish crackers. The kids couldn't stuff the crackers into their mouths fast enough.

Many parents have replicated this experiment at home. Sothern recently worked with a mother and father who routinely bought diet pop for their chubby children and ice cream for themselves. The kids would get in trouble for sneaking tastes of their parents' dessert, but how could they resist? She knows other parents who stash sugary treats under lock and key, effectively turning Twinkies into treasure.

Instead of taking a draconian, all-or-nothing approach, parents should strive for moderation, says San Francisco registered dietician Judith Levine, author of the book *Helping Your Child Lose Weight the Healthy Way*. "Start slowly," she says. "Make small changes. Pick the easiest things to change first." In other words, don't launch a nuclear attack on your refrigerator or pantry. If your child loves ice cream, buy the low-fat version or try frozen yogurt; maybe get a scoop of the full-fat version every once in a while. And don't buy the message that there's good food and bad food. "If your child likes potato chips, don't take

them away—just serve less of them," Levine says. "You can't help a person change their eating habits unless they feel good about what they're doing and happy about what they're eating. Just try to get them on the right path to making changes."

It's hard to fault parents for getting tough on food, especially when their children are severely overweight. Tracy Graham, Josh's mother (see sidebar on page 162), doesn't feel she can let her guard down for one moment. "I have to constantly reinforce him," she says. As part of her round-the-clock duties, she recently rummaged through his pants pockets and discovered receipts for french fries and donuts. "I just lost it," she says. Josh got a stern lecture, but did he really get the message?

Golan believes parents in this situation can take a different approach. Instead of putting french fries and donuts on the "enemy combatants" list and then watching for evidence of treason, parents can help children find healthier alternatives. "I would ask that child why he enjoys donuts. Maybe he can get that same feeling from something else, like a slice of bread with jam. It's not the donut that he's after—it's the feeling that donuts give."

As radical as it may sound, children can—and should—take responsibility for their own eating. Kids certainly have the capacity to make terrible choices when given the opportunity. (Remember those brownies covered with gummy bears?) But when they have a basic knowledge of nutrition, an array of healthy food options, and a few positive role models, they can do things that would make any nutritionist proud. Levine encouraged her daughter to make choices from the age of two. "What do you want for dessert?" Levine would ask her. "You can have grapes, bananas, or peaches." Then the toddler would pick her dessert "and feel like a big shot." School-age kids can help with the grocery shopping, dish up their own plates, and decide for themselves when they're full.

Molly's story: "I don't want her to be on a diet"

As a first grader, Molly Markus of Billings, Montana, put on weight seemingly overnight. In no time at all, the brown-haired girl with freckles quickly went from spunky and slender to sluggish and heavy, and her mother, Libby, was worried. "I knew she had an underactive thyroid, but the preliminary tests were fine," says Libby, a registered nurse. As her daughter kept getting larger, Libby badgered the family doctor to run more tests. He eventually—and grudgingly—referred Molly to an endocrinologist. "It was just to make me happy," Libby says. "They got sick of listening to me." Of course, she turned out to be absolutely correct. Molly was diagnosed with an underactive thyroid in second grade, and she'll have to take thyroxine for the rest of her life.

Molly's thyroid medicine gave her more energy, but it didn't magically melt off the pounds. Looking back, an underactive thyroid wasn't her only problem. She also had a bad case of the typical-American-kid lifestyle. "I'd watch TV any chance that I could," Molly says. "Mom would have to drag me away from it to get to me to daycare." And, like countless other kids, she turned the TV room into a snack bar. "It got to be that we had food jammed in the seat cushions," Libby says.

When she started fifth grade, Molly was 4'8" and weighed about 145 pounds, enough to make her stand out from the crowd. The youngster started to feel self-conscious, and her mom kept doing what mothers do—she worried. Libby took Molly to the doctor for another round of tests and got some more bad news. Molly's blood sugar was slightly above normal. Libby had to explain to her daughter that extra weight can raise a person's blood sugar and that, eventually, high blood sugar could turn into diabetes. "I felt that I either needed to lose weight or get the disease," Molly says.

As a nurse, Libby already knew more than most people about the basics of healthy living. Still, she felt like she could use some help. She signed Molly up for the Shapedown program at Billings Deaconess Hospital. Libby didn't need to lose weight herself, but she attended every session with her daughter. The program stressed the importance of physical activity and showed families how to eat in moderation. As Molly got to know other kids in the program—including a few who were much heavier than she was—she felt better about herself. "I didn't feel like I was the only person," she says. "It felt weird to know that I could have it worse."

Molly took the lessons to heart. She started spending less time in front of the TV and more time jumping on her backyard trampoline or climbing the trees in her front yard. She also took a new approach to food. "If you have a chance, you should try to eat something healthy," she says. "Don't think about fatty foods when you're hungry. Think of fruits and vegetables. It doesn't have to be greasy to be good." Libby has made it easier for her daughter to make healthy choices. She still buys ice cream, chips, candy, and other treats every once in awhile, but she

➔

doesn't keep a constant supply in the house. "I don't want her to be on a diet," Libby says. The family had plenty of chocolate lying around at Easter, but Libby says she sneaked most of it to the office the next day and gave it away. (This admission came as news to Molly. "You did?" she asked mournfully.)

Now ten years old, Molly is two inches taller and eight pounds lighter than when she started Shapedown, and her blood sugar is right where it should be. She's a bright, confident little girl who can hold her own in an adult conversation. She's talented too. In March 2004, she had a solo in the citywide girl's honor choir concert. She stood up at the microphone and belted out her number in front of a packed house and a few hundred camcorders. She wasn't a girl with a weight problem—she was a girl with a killer voice. Once the clapping started, Libby had a hard time stopping. •

Nobody is saying that ten-year-olds should be given carte blanche in the grocery store, in the restaurant, or at home. Parents are still the last line of defense between children and the temptations of the world. Parents may need to take a couple of steps back, but they can't leave the picture entirely. In fact, moms and dads are left with a daunting task: in a world of junk food and quick calories, they have to work hard to create an atmosphere that encourages healthy choices. As Fisher puts it, "Foods are grotesquely abundant. Parents need to make decisions to set up the environment in such a way that there's not a constant struggle."

If you opened Susan Ryals's refrigerator a few years ago, you might have found a gallon of whole milk, a plate of leftover fried chicken with a side of macaroni and cheese, and a two-liter bottle of Coke. These days, the shelves are more likely to hold a gallon of skim milk, a huge bowl of fruit salad, a plate of leftover stirfry, and several cans of Diet Mountain Dew. (Susan says she goes through a twelve-pack of Diet Dew every day, making her one of the most highly caffeinated people in Choctaw County. Amber prefers bottled water.)

Amber and her sister still have potato chips, ice cream, and chocolate every once in awhile, but not every day. "I never tell them what to eat," Susan says. She trusts that her girls have the knowledge and the good sense to make the most of their options, and they're proving her right.

Susan Ryals also takes a "hands-off" approach to weight. Even though she keeps a scale in her kitchen and checks herself daily, she never asks Amber about her weight, and she certainly never suggests that her daughter should lose a few more pounds. She's no expert in child psychology, but she seems to have found exactly the right approach. Golan believes parents who stress dieting and weight loss put undue pressure on kids. Instead of telling their kids to "get thin," parents should encourage them to "eat healthy."

A renewed commitment to the family meal is one of the most important gifts we can give kids. A recent study of more than 16,000 boys and girls between the ages of nine and fourteen found that kids who regularly ate at home with their families had the healthiest diets—by far. Compared with kids who usually ate alone or grabbed something out of the house, kids who had regular family meals had less fat in their diet but got more fiber, iron, calcium, folic acid, and vitamins C, E, B6, and B12.

Kids may protest that they'd rather watch television instead of listening to their parents talk about their day, but as it turns out, they actually crave that time with their families. In 1993, Oprah Winfrey conducted a "Family Dinner Experiment" in which five families volunteered to eat dinner together every night for a month and keep journals on the experience. Several families reported that at first it was a chore and the children found it tedious. But afterward, when the families appeared on Oprah's show, the parents were amazed to learn how much their children treasured that time at the table.

If you're at all concerned about our fast-food culture, the family dinner table is the perfect place to take a stand. You and your family can choose the ingredients in your meals and make sure everything is

Healthy Eating at Every Age

So what do we mean when we tell a child to eat healthy? In this age of fast food, vending machines, and carpet-bombing marketing, it's easy to lose sight of what kids really need. Here's a guide to feeding kids of every age.

Nutrition for Infants

New mothers have heard the message for many years that breast milk is best for their babies. At a time when far too many kids are growing fat, the message takes on a whole new urgency. In addition to all of the other benefits of breast milk—disease-fighting antibodies; easily digested proteins; and easily absorbed iron, calcium, and other nutrients—breastfeeding may help lower the risk of childhood obesity years down the road. Researchers speculate that breastfed infants have more control over their own calorie intake. Breastfeeding simply isn't an option for some moms, but those who can do it will be giving their babies the best possible start.

At four to six months, most babies start to show signs that they are ready for solid foods. They may start to express interest in food by making chewing noises or reaching for your plate. In general, it's safe to start trying solid foods when babies weigh at least thirteen pounds or have doubled their birthweight and can sit up with a little bit of help.

Most experts recommend starting with iron-fortified rice cereal, which is easily tolerated and helps babies to meet their iron needs. Then, parents can add one new food at a time, usually single puréed vegetables and fruits and then strained meats and poultry. Some babies may be sensitive to common allergens like peanuts, soy, eggs, fish, shellfish, wheat and milk, so introduce one new food at a time. Better yet, save egg whites and nuts until after the first birthday.

As teeth start coming in, babies can handle more "finger foods" such as teething biscuits and cut-up ripe fruits and cooked vegetables. In addition to adding more variety to the diet, this offers an opportunity for developing eye-hand coordination. Of course, as soon as you start giving your child substantial foods, you have to be very careful about choking. The top choking culprits for young children include hard candy, popcorn, pretzels, chips, spoonfuls of peanut butter, nuts and seeds, hot dogs, raw carrots, raisins, and whole grapes. The American Academy of Pediatrics urges parents not to serve these foods until a child turns four.

When feeding your baby, aim for variety in tastes, colors, and textures. Trying and enjoying new foods will give your baby a solid foundation for a balanced diet later in life. Many parents manage to break through the built-in distrust of new foods with baby-approved fruits and vegetables such as applesauce, pears, plums, bananas, sweet potatoes, carrots, squash, peas (without the pod), and green beans. (All of the vegetables have to be well cooked.) ➔

Most babies are ready for a trial with a sippy cup around six to nine months, but don't be too generous with the juice. Even one hundred percent juice is loaded with sugar. The American Academy of Pediatrics recommends holding off on juice until six months of age and limiting daily intake to six ounces.

Even if you're concerned about your chunky infant growing into a chubby child, this is not the time to give your child skim milk or nonfat cheese. Babies need fat! The extra calories help them grow, and fatty acids are crucial for the development of the brain and nervous system.

Nutrition for Toddlers and Grade Schoolers

After the first year's rapid spurts, growth now starts to slow down, and a child's appetite may flag. It's not uncommon for parents to become alarmed if their child seems to eat less day-to-day. Parents should try to look at the big picture rather than a snapshot of a child's diet. Children tend to eat as much as they need over the course of a few days.

At the age of two, children should switch from whole milk to one percent or skim milk. Whole milk is the leading source of saturated fat in kids' diets, and it's completely unnecessary. Kids will easily adjust to low-fat or nonfat milk, especially if parents don't make a big production about the switch. Water is another great beverage for kids. One hundred percent fruit juice is all right in small amounts—about eight to twelve ounces every day. Soda, Kool-Aid, and other sugary drinks shouldn't be everyday beverages.

In addition to providing vitamin D and calcium, low-fat milk gives growing children a much-needed dose of high-quality protein. The protein supply can be enriched even more with eggs (especially egg whites), beans, peas, and lean cuts of meat.

Children need plenty of high-fiber foods, such as fruits, vegetables, beans, and whole grains and cereals. These foods will satisfy their appetite without a lot of fat and calories. When picking out produce, look for color, especially dark green, deep yellow, and orange. These colors signal rich supplies of vitamins A and C, potent disease-preventing antioxidants.

Snacks are important for people with small stomachs. Kids usually need a little something to tide them over from one meal to the next. Choose healthy snacks, such as fresh or canned fruits and vegetables, yogurt, cereal, low-fat or skim milk, low-fat cheeses, lean meats, rice cakes, or graham crackers.

The Teen Years

As kids grow older, they make more of their own food choices, and good nutrition isn't necessarily their top priority. Not only do teens tend to go overboard on fats and sugars, they also fall short of several key nutrients. Surveys by the USDA show that teenage girls typically get less than eighty percent of the recom-

➔

mended daily allowance (RDA) of magnesium and calcium, and both boys and girls tend not to get enough zinc and iron. Here are some foods that can fill the gaps.

Calcium: Low-fat dairy foods such as one percent or skim milk, yogurt, or cheese; calcium-fortified beverages such as one hundred percent fruit juice and soy milk; dark green vegetables, especially the leafy kind such as kale, collard greens, bok choy, and broccoli; and tofu made with calcium sulfate.

Magnesium: Whole wheat breads and cereals, bran cereals, black beans, black-eyed peas, spinach, pecans, and peanut butter.

Zinc: Beef, enriched cereals, wheat germ and wheat bran, oysters, crabmeat, black-eyed peas, sunflower seeds, and almonds.

Iron: Beef, chicken, pork, dried fruit, green leafy vegetables, iron-fortified cereals, beans, spinach, pumpkin seeds. (Iron is more easily absorbed when it comes from an animal source.) •

cooked with care. (Who knows what's going on in a restaurant kitchen or a food factory?) Eating becomes a group project, and everyone benefits—especially kids.

Forty years ago, the typical family meal was a two-hour project. Nowadays, fifteen minutes can seem like a big commitment. The good news is that a healthy family meal can fit into just about anyone's schedule.

For starters, you should take a few minutes to plan your meals for the week before heading to the grocery store. Make a list of the ingredients that you'll need, and hold onto the list so you can use it for future trips. It's much easier to put together a meal in a hurry if you don't have to make a special trip back to the store. When you have a little extra time to make a meal, prepare more than one batch. Many meals—such as soups or casseroles—keep well in the freezer. Invest in a crock pot—it does the cooking while you're at work. Grilling, stir-frying, broiling, and microwaving all take less time than baking. Serve easy-to-prepare breakfast foods at dinner, such as omelets, or add dried fruit and nuts to a bowl of hot cereal.

Grocery stores and food companies have certainly taken notice of the modern time crunch, and they've responded with a lot of timesaving food options. When planning your week's menu, consider building meals around these convenient items.

- Rotisserie chicken (leftovers are great for soup or chicken salad)
- Skinless chicken strips
- Precut vegetables to eat raw or toss into a soup or stirfry
- Prewashed salad mixes and spinach
- Frozen vegetables

Here are some more big and little ways parents can help children of all ages become healthy eaters.

Examine your feelings about food. Our kids absorb our attitudes whether we speak them out loud or not. If you believe that the only way to lose weight is to go hungry or forever give up your favorite foods, your children will incorporate those beliefs into their own way of thinking. Your child could decide that weight loss can be achieved only by suffering, and in that case he'll just stay the way he is. If, however, you take a moderate, supportive approach to eating right, he will be more likely to participate willingly.

Shop smart and shop together. Learn to read labels for fat, sodium, and sugar content, then teach your kids to be label readers too. Ask your child to help you write the shopping list, and bring him with you to the grocery store, where he can choose some foods—fruits, vegetables, low-sugar cereals, and yogurt, for example—that he'll eat happily later on. If you buy high-calorie treats, buy enough just for the day that you're shopping.

How to Read a Food Label

For all of the excellent books in the world about childhood nutrition, some of the most important reading parents can do is printed on food labels. All parents should become avid label readers and teach their children to do the same. The ingredient list on breakfast cereals is a great place to start. Parents can explain that the food contains more of the first ingredient than anything else. If a child is old enough to read the word "sugar," she's old enough to know that Cookie Crisp and Apple Jacks aren't the best choice.

The ingredients list on cereals is also a good place for more advanced label readers to hone their skills. The first ingredient listed for "Magically Delicious" Lucky Charms is whole grain oats, a very promising start. But things go downhill from there. Different types of sugar—including plain sugar (sucrose), dextrose, and corn syrup—come up five times in the list. All told, sweeteners makes up nearly half of the weight of the cereal. It's debatable whether they're delicious, but magical they are not.

Food labels really aren't too hard to decipher. Here's a quick guide to the fine print:

First stop: the serving size

You'll see both the serving size, which is based on common household measurements, and the number of servings in the package. Beware! The serving size on the package may be different from what you actually eat. If you eat twice the serving size, you must double all of the nutrition facts too.

Count the calories

Calories are listed per serving. You'll notice that the label tells you how many of those calories are from fat. Ideally, no more than thirty percent of your total calories should come from fat, and no more than ten percent should come from saturated fat. This guideline leaves plenty of room for fattier foods—just be sure to balance your day out with leaner foods.

Next come the nutrients

When reading food labels, remember that nutrients come in two different categories: some to go easy on and some to seek out. Labels list the potential troublemakers first, starting with total fat then moving down to saturated fat, cholesterol, and sodium. Excessive amounts of these nutrients can raise the risk of heart disease, high blood pressure, and some cancers. A quick glance will show you the number of grams of each nutrient, and more important the *percent daily value,* written as *%DV.*

The percent daily value is the key number for gauging a food. If a label says that the frozen lasagna you're considering for lunch will give you ninety percent of your total fat and seventy percent of your saturated fat for the day, you should

➔

either find a different option or plan a practically fat-free dinner. Keep in mind, however, that %DV is calculated on a 2000-calorie diet. If you happen to be an elite athlete who needs 4000 calories a day, you'll have to mentally cut all of those figures in half. If you're choosing foods for a child who gets less than 2000 calories a day, you'll have to adjust those %DVs upward.

By 2006, labels will also list the amount of trans fatty acids, also known as trans fats. This stealth fat—often found in pastries, donuts, store-bought cookies, stick margarine, and french fries—is harder on your arteries than any other kinds of fat. Until labels start listing trans fats, you should do a little sleuthing. If the label lists hydrogenated or partially hydrogenated vegetable oil as a major ingredient, the product is bound to be rich in trans fats. When possible, choose foods made with naturally unhydrogenated oils such as canola or olive oil. Contrary to common belief, peanut butter does not contain a significant amount of trans fats.

When you buy foods specifically intended for children under age two, the labels may not list the calories from fat or saturated fat, the saturated fat, or the cholesterol content. Because kids under two shouldn't be on a low-fat diet anyway, this information isn't really important.

Of course, labels list good guys too, specifically vitamins A and C, iron, and calcium. All of these nutrients are associated with good health. Again, the quantity of these nutrients is expressed as a %DV. Generally, a food that provides five percent or less of the daily value is considered low, and twenty percent or higher is high. You can also use the %DV to compare one product with another.

Tackling total carbohydrate

By current USDA standards, roughly sixty percent of your calories should come from carbohydrates, including starches and sugars. For a person getting 2,000 calories a day, that works out to about 300 grams of carbs a day. The %DV for total carbohydrates will show you how close a food takes you to that goal.

Pondering protein

Usually, you'll just see the number of grams of protein listed on the label. A %DV is only required if the product has a claim such as "high in protein." Food intended for infants and children under the age of four must also list the %DV for protein. •

Make sure you know what fruits, vegetables, and lower-fat foods appeal to your kids. Levine says that when she asks kids to name foods they like, their parents are sometimes surprised about what ends up on the list.

Let your child decide when he's full. Children—especially young children—generally know when it's time to stop eating. As they go through growth spurts and lags, their calorie needs will change from time to time. It's important for parents to anticipate these changes, not "force feed." Start with age-appropriate serving sizes and let your child ask for more if he's hungry.

Don't skip breakfast. Just like adults, kids need to start off their day with a good meal. Breakfast gives kids much-needed energy, and it keeps them from overeating at lunch.

Put food in its proper place. Teach your children that food helps them grow and stay healthy. Using food to reward or punish your child may lead to behavior problems related to food and mealtimes.

Keep a schedule. Children thrive on routines. A regular dinner time is just as crucial as a regular bed time. Not only are children happier on a schedule, it also gives them a chance to maintain their normal hunger cues.

Help your kids expand their culinary horizons. You'll get a surefire "no" if you ask a young child whether she wants to try something new. Instead, put a bit of the new food on her plate. Your child may genuinely not like everything you serve. If this is the case, offer it again a few weeks later or prepare it differently. It helps, too, if you don't behave as though you're eating your vegetables in a forced march toward good nutrition. Try perking up those carrots and green beans with spices or lemon juice. Or cut up some fresh vegetables and serve them with a low-fat dip. You might also try serving the vegetables as a first course so your kids will gobble them up at the beginning of the meal when they're really hungry.

You might have to offer new foods five or even ten times before your youngster digs in. Unfortunately, it usually doesn't take much coaxing to get a two- or three-year-old to try a hamburger or french fries. If kids get too much fatty, high-calorie food early in life, they can develop a lifelong love affair with these foods that will be nearly impossible to break.

It's all right to serve the same food several meals in a row. It's not uncommon for children to go on food jags where they want peanut butter sandwiches or spaghetti over and over again. Wait to offer new foods until your child is rested and happy and offer them at the beginning of the meal. Offer a small amount of other foods and encourage your child to try them. If you are not successful, remind yourself that this food jag will pass. Remain calm and try not to let your child sense your frustration.

Encourage children to eat slowly. It takes approximately twenty minutes for the brain to signal that the body is getting food.

Introduce vegetables early, as soon as your baby is eating solid foods. Offering only puréed fruits might lead him to believe that all foods should be sweet. The rule of thumb for vegetables should be yellow first, orange and pale green next, dark green and red last. In fact, wait until the end of the first year to introduce baby to beets, turnips, spinach, mustard, and collard greens—before that, his stomach may not be equipped to handle their high nitrate content.

Make healthy substitutions. Many recipes can handle a tweak in fat or sugar here or there without anyone noticing. If you're baking, for example, you can reduce a cup of oil to three-fourths or two-thirds of a cup instead. For sweet breads such as banana bread, you can cut the oil in half and replace the rest with applesauce. You can also replace half the whole eggs in a recipe with egg whites. (For a recipe that calls for two whole eggs, you can substitute one egg and two egg whites.) Instead of whole milk, try nonfat or one percent milk. Reduced-fat cream cheese and sour cream can often go undetected, especially in baking.

Try switching to healthier fats or eliminating some fats altogether. The kids in the Shapedown program learn that they can skip the butter on their toast in the morning and instead use a fruit spread without sugar. Try substituting healthy oils such as olive, canola, soybean, sunflower, and peanut oil for lard, butter, and palm oil. Ideas for low-fat cooking abound these days. You could buy a low-fat cookbook, borrow one from the library, or search the Internet for more ideas.

Karen Rodgers has discovered firsthand that small steps add up. When she found out how much sugar is packed away in a fruit roll and how much fat lurks in some of the ostensibly healthy granola bars, she switched Jake's snacks to fresh fruit or pretzels. "I've tried to teach him about healthier things and to be creative. Now I broil meat or barbecue and use the least amount of fat possible," she says. These days, instead of buttering Jake's rice, Karen flavors it with chicken broth and garlic.

Getting Kids Moving

As a rule, good nutrition alone isn't enough to help anyone lose weight and keep it off. Whether you're thirteen or thirty, exercise has to be a part of the equation. Of course, staying active isn't always easy. Imagine being a large girl in, say, Butler, Alabama. You live too far from school to walk or ride your bike, and your town doesn't have a decent park, let alone a gym or a swimming pool. So what do you do?

Amber Kearley had every reason to stay on the sidelines, but, with her mother's encouragement, she decided to get in the game. As a sixth grader, Amber bravely squeezed her 200-pound body into a basketball uniform and joined the school team. Even though she always ran hard in practice (hard, not fast), she almost never got off the bench during games. And then came the nutrition classes, the changes around the house, and the regular walks with her mom. As the weight started to come off, Amber reached new heights of speed and stamina and started getting more and more playing time. Her varsity team made it all the way to the state "Final Four" in 2004, and Amber was right in the middle of the action.

When the world conspires to keep kids sitting on the couch or glued to a screen, parents must do their part to encourage physical activity. Once again, it all comes down to the environment that parents build for children. Whether they live in Butler, Seattle, or the inner city of Chicago, parents can create an atmosphere that makes children want to move.

Here are some ideas to help you keep your kids active.

Tap into your child's penchant for play. "Set up a soft mat in the den, put a swing set or some balls in the backyard, put some trucks or a box of dress-up clothes in the playroom—anything to get them away from the TV," Sothern says. For older kids, try setting up a basketball hoop in the driveway or a trampoline in the backyard. You might even want to buy a secondhand stationary bike in a yard sale and set it up in the living room—that way you or your kids can choose to pedal while you watch TV.

Be a role model. When kids see their parents going for walks, riding bikes, or shooting hoops, the lesson sinks in. A study from the early 1990s found that kids are six times as likely to be active if their parents are active. The best way to set an example for your kids is to do things together. Instead of watching TV together before dinner, if possible, use that time to play outside with your kids. Take a nature hike, throw the football around, sign up for a parent/kid aerobics class . . . whatever sounds fun.

Use rewards. James Sallis of San Diego State University says kids will gladly exercise if they see a payoff. Tell them they can watch TV after they've walked around the block a couple of times. Thirty minutes on a bicycle might be worth a trip to the mall.

Get outside on the weekends. Try not to let work or chores take over your weekends and ruin your chance to relax with your family. Plan special family outings such as hiking or rollerblading in a nearby park. If the parks in your neighborhood aren't very inviting, drive or take the bus to the closest park that's clean and safe. Take a (healthy) picnic lunch with you—that way you can have hours of fun without straining your wallet.

Walk to school with your child. This is one of the best ways to get some near-daily exercise with your kids. For support from like-minded parents and community members, you may want to check out Safe Routes to School, Walk and Bike to School, and Walking School Bus.

Do your errands on foot. Whenever possible, walk to the store instead of driving. Visit your neighbors on foot. Take public transportation whenever possible. Show your kids that you don't always need wheels to get around.

Enroll your child in a sports or dance class. Choose something that he or she really enjoys, such as martial arts, dance, gymnastics, tennis, or swimming. Make sure it's a class your child is interested in, not necessarily the one you dreamed of taking as a child. Check out the local YMCAs and community centers for free or low-cost swimming lessons, scholarships, and summer camps. If your child has an interest in sports, encourage him or her to try out for a school team. If your school has eliminated sports or physical education, contact your legislator about the possibility of restoring them.

Assign active chores to every member of the family. Rotate the chores—such as vacuuming the house or washing the car—to help ward off boredom. Perhaps the most interesting "chore" would be to help plant, tend, and harvest a family garden.

Don't set your expectations too high. It can be hard for heavy children to get moving. For one thing, exercise may make them feel self-conscious. But even more important, it can bring on extreme discomfort and physical pain. Overweight children have a much lower "exercise tolerance" than other kids, Sothern says. After five minutes of strenuous exercise—about the time most kids start to get out of breath—hefty kids may be hitting the wall. Their faces turn red, their breathing becomes extremely rapid, and lactic acid builds up in their muscles, causing intense pain.

Encourage your overweight child to exercise slowly and steadily, especially at the beginning. Extremely obese kids can swim slowly or ride an exercise bicycle with the resistance turned down to zero for twenty minutes twice a week. Kids who are only slightly overweight might try brisk walking, inline skating, martial arts, or other weight-bearing exercises for twenty to thirty-five minutes three times a week. Over time, workouts can become longer and more intense.

Keep the fun in sports. A lot of kids naturally gravitate toward sports, then find a way to drop off the team when it's no longer enjoyable. Parents should be supportive cheerleaders, not critics of their kids' performance. Children who feel pressured or criticized—or who really get yelled at by a parent—will walk away from sports. If your child is inclined toward a team sport, find a coach who's supportive of the kids. In the summertime especially, water sports such as swimming or even playing Marco Polo in the local pool, lake, or seaside can allow your child to burn a lot of calories while enjoying the day.

On a rainy day, put on a fitness video that encourages kids to sing, jump up and down, or learn new dance moves. If you and the kids are going stir-crazy, take a trip to the local bowling alley, skating rink, or climbing gym.

Cut back on TV time. The "off" button on the remote can be one of your most powerful tools as a parent. Limit your child's TV time, and if there's a television in your child's bedroom, take it out. Research shows that a TV in a child's bedroom is a strong risk factor for being overweight, even among preschoolers. Even if you never try to improve your family's meals or take a single family walk, simply turning off the TV can boost your child's health—and trim his waistline. In 1999 researchers at Stanford University convinced a group of kids to cut their TV time by one-third to one-fourth. The researchers told the kids not to change their diet or exercise level. After one year, the kids had gained significantly less body fat than a control group of other kids their age. They also had smaller waist sizes and better waist-to-hip ratios and body mass indexes. If strangers in white coats can have such a positive effect, imagine what a caring parent can do.

Karen Rodgers first cut Jake's television watching down to twenty hours a week. A few weeks later, he was down to thirteen hours. Now Jake is lobbying for a treadmill so he can exercise and watch TV at the same time. (In the Shapedown program, if you're working out while watching television, the TV time doesn't count as part of your allotted hours.)

Along with paring back his TV time, Karen has established a new rule: Jake can't eat while he's watching television. "A lot of times I'm not good at enforcing that, because I'm trying to do laundry or get other things done around the house. I feel bad about that," she says. As a single working mom, Karen feels most vulnerable to giving in to the tube at the end of a long day. "That's when I have to tell myself, 'You need to get off your butt, and you need to take Jake for a walk.'"

Back in Butler, Amber has plenty of chances to walk. She walks around the abandoned clothes factory with her mom. On weekends, she walks with her friends at the mall in Meridian, Mississippi. Most of all, she walks down the halls of her school with newfound pride. Her friends often ask her how she managed to lose so much weight. Like so many other kids—and adults—in the world, they're hoping to find the "secret" that will quickly solve their problem. But what can Amber really tell them? They should exercise more. They should pay attention to what they eat. Most of all, they should have parents like Susan Ryals. With someone like her at their side, almost anything is possible.

8
A New Deal for Kids

THIRTY YEARS AGO, new fathers thought nothing about lighting up a celebratory cigar right outside the hospital nursery. People routinely smoked during movies, whether they were watching *The Godfather* or the matinee showing of *Herbie the Love Bug*. Smoking was commonplace in grocery stores, department stores, fast food restaurants, teachers' lounges, and, of course, airplanes. Some universities allowed students to smoke during class, and many high schools set aside special smoking areas for kids. Looking back, it all seems incredible. Why did we ever think it was okay to surround kids with cigarette smoke?

Our beliefs about what is acceptable for children today will resonate for decades to come. Twenty years from now, we may well look back on the current generation's steady diet of soda, fast food, and television with the same horror with which we now view cigarette smoke in a nursery. By then, the full consequences of the obesity epidemic will be in plain view. Many of today's heavy kids will have grown up to be obese adults, and a few will have already had their first heart attack or diabetes-related amputation.

Armed with information about tobacco's dangers, we changed our attitudes about cigarette smoking. Now we face a question for a new era: what are we going to do about our fattening culture?

For now, there's no shortage of talk. In June 2004, scientists, health experts, and government officials gathered in Williamsburg, Virginia, for a national obesity summit. Health and Human Services Secretary Tommy Thompson used the occasion to tout his Small Steps campaign, an effort to get people to make small lifestyle changes. "First, we have to work hard to spread the gospel of personal responsibility," Thompson said.

"We're not asking anyone to run a marathon, join a gym, or give up eating," Thompson continued. "We're talking about small steps—play outside with your children, snack on fruits and vegetables, take the stairs instead of the elevator. These small steps can really make a big difference in our health." So they can. But they aren't the stuff of a serious national campaign to end one of the worst epidemics in human history. Thompson's main message seemed to be that government is there to act as a cheerleader, not a leader.

The health department's biggest push so far has been a bizarre television ad campaign featuring flabby body parts that have been "lost." In one ad, two kids find a disembodied belly on a beach and poke at it with a stick. The ads are intended to encourage viewers to lose their fat, although typical viewers might be much more likely to lose their lunch.

State governments generally haven't distinguished themselves on the obesity issue either, according to a national obesity report card from the University of Baltimore issued in August 2004. When the report card rated states' efforts to control the obesity epidemic, twenty-three states failed. Attempts to manage the epidemic in children were slightly more vigorous: on that score only eighteen states failed. Not one state received an A. "Simply put, state governments are not addressing this problem effectively, and [the neglect] is doing a lot of unnecessary damage," said Zoltan Acs, a professor of economics and entrepreneurship at University of Baltimore and an author of the study.

Thompson's federal department has repeatedly offered only "encouragement" to the food industry to make changes, not the force of law. At the summit, Thompson lauded voluntary food industry changes, including Kraft Foods' announcement that it would stop printing up posters and book covers with its logo for use in schools and McDonald's decision to offer healthier options. "We sounded the alarm, and the free market is responding," he said. In other words, industry is cooperating in the battle against obesity, and the department of health is satisfied that Big Food is self-regulating without disruptive limits.

But Yale University's Kelly Brownell and other participants at the national obesity summit say this is nonsense. "The early days of tobacco history involved everybody standing around the campfire holding hands and saying we need to collaborate, cooperate," Brownell said. "We paid a huge price for that because it stalled public health efforts to control tobacco for decades, and who knows how many millions of people died as a consequence."

Former surgeon general David Satcher says he doesn't put much faith in small steps either. "I have a lot of respect for Secretary Thompson, [but] I disagree with that as I've disagreed with some other things," he told reporters at a Casey Journalism Center conference in March 2004. "I think we need some big steps to deal with this. This is a very serious epidemic. A lot of people are going to die because of this epidemic. It is time for serious steps, big steps. . . . I don't think any epidemic can be stopped without major public commitment."

Still Waiting

Those on the front lines are especially disappointed by the dissonance between the government's rhetoric and its actual response. "The anthrax scare may have affected six individuals, and it received an enormous amount of national attention," says physician and diabetes researcher Roberto Treviño. "Within months we had new treatments, vaccines, and a diagnostic test." In San Antonio alone, he says, "We expect 5,000 youth to go blind, lose limbs, or be on dialysis because of diabetes, but our society does not give them the needed attention."

If our leaders are serious about doing something to curb childhood obesity, they would be wise to start treating it like the national emergency they say it is. Instead of showing us disembodied stomachs on the beach and posting helpful hints for grocery shopping on websites, our leaders in Washington could create a high-level government office to coordinate the campaign against obesity.

The CDC has certainly tried to tackle the problem, but its recommendations—most notably those on physical activity—have gone largely unheeded. In a report to President Clinton in fall 2000, the CDC repeated a prescription it has written time and time again: "Quality, daily physical education" should be available for all children from prekindergarten through twelfth grade. Furthermore, all schools should have certified physical education specialists; appropriate class sizes; and "the facilities, equipment, and supplies needed to deliver quality, daily physical education." But instead of mandating recess and PE in every public school, the White House and Congress have been silent on the subject as schools cut physical education, eliminate trained PE teachers, consolidate classes into groups of eighty kids or more, and sell Coke to buy gym equipment.

The CDC also made it clear that the country needs after-school programs with trained coaches and recreation program staff, plus a host of community design features meant to boost physical activity—well-maintained sidewalks, bicycle paths, trails, parks, and recreation centers. But the reality is that most cities can't pay for these necessities, and as states and cities struggle with their budgets, community swimming pools and recreation centers are closing amid howls of opposition from kids and parents. No amount of small steps will fill these giant gaps in U.S. schools and neighborhoods.

What we need is a New Deal for Kids—radical steps to fight obesity on a national scale, much like Franklin Delano Roosevelt's blueprint for putting people back to work during the Great Depression. The federal government should take action immediately to help end this epidemic.

For starters, Congress should plug the loophole in USDA policy that allows junk food and soda companies to skirt the regulations on healthy food at schools. We need a national nutrition policy in schools that would regulate soda, foods sold in vending machines, and other foods that compete with the school lunch program. This could be modeled on the strict Texas policy for competitive foods in schools, which bans the sale of competitive foods during lunchtime, forbids kids from buying vending machine foods and soda until after school, and limits soda to twelve-ounce cans. Implemented nationwide, such a policy would ultimately safeguard kids from unhealthy foods while giving food service directors the financial wherewithal to keep improving their meals instead of being locked in competition with Pizza Hut and Taco Bell on their own campuses.

As part of the New Deal, public education and all of the government's mandates should be fully funded. Schools shouldn't be forced to raise capital from fast food to pay for PE and other necessities. If they can't afford paper, pencils, and textbooks, schools are unlikely to be able to offer PE and recess to the kids who most desperately need it.

The ultimate solution to the school food issue does lie at the federal level, says Margo Wootan of the Center for Science in the Public Interest. "If we're going to have nutritional standards in place for the school meal program, we should also have nutritional standards in place for other foods sold in schools," Wootan says. "We need federal leadership. We have been working to get Congress to give the USDA the authority to regulate the sale and availability of all foods on campus through the school day. It's such an obvious solution."

No advocates of better nutrition in schools are holding their breath waiting for Congress to turn against the food industry and its cadre of lobbyists. But Wootan and others say that bills setting standards for competitive foods are picking up steam from legislators increasingly alarmed by the sheer numbers and the health problems of overweight kids.

A Million Mad Moms

Little of significance has ever changed in this country without a groundswell of protest. Segregation, the Vietnam War, discrimination against disabled people: the impetus to end all these came not from government but from the people themselves. Parents and others who want to stop obesity in kids need a national movement that will overwhelm the toxic environment. "Give me a million mad moms," said Texas Agricultural Commissioner Susan Combs, "and I'll get you some organization at school." Parents need an organized way to make their voices heard, whether it's to protest or support a piece of legislation, to fund-raise for candidates, or to push for school reforms. Like current social activists, members of this movement could take advantage of the immediacy of email and the Internet. Consider the possibilities if such an organized group were to flood Congress with letters, faxes, and emails with the same ferocity as Big Food.

If parents can do that on a small scale, as they did with Aptos Middle School, they can influence change on a national level as well. Parents across the country, in their own communities, have had success limiting soda and junk food in schools. Wootan and others say that sometimes it just takes one or two parents approaching a school board or school superintendent to get the ball rolling.

To start, parents and school administrators could simply replace the ubiquitous bake sale with a healthier and more fiscally rewarding alternative. Parents at a Los Angeles elementary school got rid of their bake sale after one parent complained that it was completely counterproductive. Once a week for the entire school year, busy parents who don't have time to bake were purchasing baked goods to sell, only to have to turn around and buy someone else's cakes or cookies for the cause. Instead, parents now host a yearly "fun run" that gets their kids moving and raises more money than the bake sales ever did. "A lot of times it takes someone in the community—a dentist or a mom or a health teacher—to really shake things up and say, 'What are we doing?'" Wootan says.

Schools could also become a hub for federally funded obesity screening. The International Association for the Study of Obesity notes that schools in Singapore and Hong Kong track the weights of students. Like immunizations or vision check-ups, obesity screening would become universal. Some experts are also floating the idea of diabetes screening for high-risk kids and teens—low-income minority children who have a family history of the disease.

Treviño, who works closely with that population, would favor diabetes screening only if kids with high blood sugar were sent to prevention programs that included medical treatment and help with changing their environment and unhealthy behaviors. We cannot screen children and send back the kids who test positive "with only a slip of paper," Treviño says. Francine R. Kaufman, MD, head of the Center for Diabetes and Endocrinology at Children's Hospital in Los Angeles, agrees that children screened for obesity also must have follow-up care. "If [schools are] going to help them hook up with a healthcare provider, then that's a great idea. If there's a school-based health clinic, that's fantastic," says Kaufman. "But if they send a note home and that kid has no (doctor to go to), we haven't done that kid one iota of good."

Unfortunately, in the current climate, getting referred to a doctor doesn't necessarily guarantee success—or even access to treatment. When it comes to obesity in kids, many doctors feel frustrated on several fronts. For one thing, they may lack the training and nutrition education to deal with overweight children and adolescents. A study by researchers at Rhode Island Hospital found that one-quarter of southern New England pediatricians thought that they were not at all or only slightly competent to address their patients' weight issues. Some twenty percent of the doctors said they were very uncomfortable talking to young patients about their weight. In a separate study, when a researcher at the University of Alabama tested the nutrition knowledge of doctors and patients at a family practice clinic, seven percent of patients scored higher than the average physician. Now some teaching hospitals are calling for nutrition education to be a larger part of every physician's education.

But even doctors who feel equipped to treat overweight kids are regularly thwarted by the health insurers who won't cover obesity until it results in disease. Treviño and many other doctors believe the country's insurers need to wake up to the fact that it would save them money—and protect their patients from needless agony—if they paid for obesity prevention, or even reimbursed doctors for treating obese patients. "I'll say to them, 'Can I send the patient to a nutritionist?' And they'll say, 'No, we don't reimburse for a nutritionist.' So I'll say, 'Can you give me a code for obesity?' 'No, we don't reimburse for obesity,'" Treviño says. "It's ironic, but Medicaid will pay me $60,000 a year for a dialysis patient, but not $100 to develop a child's exercise program."

Fixing Corporate Culture

Many of the working parents whom author Jill Andresky Fraser interviewed for *White-Collar Sweatshop* said they felt torn and guilty about how unavailable they were to their sons and daughters. Fraser writes, "Although a host of youth crises riveted public attention by the turn of the millennium—declining educational achievements, illicit drug use, underage crime, bizarre acts of violence—there was little recognition of the role that insatiable employers had played in eroding the bonds between parents and children, decreasing family stability, and damaging the nation's home life."

Indeed, one group has so far escaped its share of the blame in the obesity epidemic—corporations: not just the companies that make fast food and soda and lobby forcefully against industry regulation, but the ones that seemingly have nothing to do with the issue. Modern workplaces encourage overwork and long hours to the detriment of their employees' families. Not only are American parents working longer hours, but nearly seventy percent of them say they don't have enough time for their kids, according to a survey conducted by the Families and Work Institute. It stands to reason that these overworked parents would have trouble keeping track of what their kids are eating and how much exercise they're getting, let alone being able to sit down to a regular family meal.

Health Secretary Thompson says corporations can be instrumental in fighting this epidemic by offering such amenities as onsite gyms. U.S. companies are to be lauded for setting up treadmills, but they need to understand that the health of our country depends on workers having a home life. Kids shouldn't have to live on take-out while their parents trudge home at 8:30 at night or miss out on family outings because the adults are working every weekend. If our kids are to recover from our fattening culture, the nature of work and career has to change. Just as they developed policies to curb insider trading, sexual harassment, and discrimination, corporations should take the lead in creating work environments that make it possible for their employees simply to be able to sit down to dinner with their families.

Not Bowling Alone

There's no shortage of families in isolation as people work harder and longer hours and kids play outside on their own less and less. Communities, in fact, are clearly part of the solution to the obesity epidemic in kids. "Having a sense of comfort with your geographic community is important," says San Antonio health collaborative director Joan Miller. She recalls a comment a psychiatrist made right after September 11, 2001—that a community walking program is just what people need to feel connected. "If you're walking in your neighborhood and you see familiar faces, then your environment doesn't seem so strange. There's a level of familiarity, comfort, security," she says. "You're riding your bike or walking, and you smile and wave, and it gives you a feeling of 'We're all in this together.'"

Whether it's through after-school exercise programs or community- and school-based gardens, the real work against this epidemic will likely have to been done at the grassroots level. And in low-income neighborhoods, that means finding innovative ways to give people access to healthy food.

In West Oakland, an economically depressed section of the city across the bay from San Francisco, a small nonprofit sells organic fruits and vegetables the same way kids used to get ice cream on a hot summer evening. The People's Grocery Mobile Market is an orange and purple truck that blasts reggae as it makes the rounds of the neighborhoods. Parents buy everything from collard greens and sweet potatoes to peppers and garlic, while kids come running up to call out their orders for peaches, plums, cherries, and other fruits. "They buy apples too," says Brahm Ahmadi, People's Grocery cofounder, "but the other fruits are more rare here, so they're very excited to get them."

Only one supermarket serves West Oakland's 25,000 people, but there are thirty-six convenience and liquor stores, only three of which sell fresh produce. Residents without cars are captive to these corner stores, where prices are thirty to one hundred percent higher than those at supermarkets. The Mobile Market gives the people in West Oakland a chance to buy seasonal fruits and vegetables at a reasonable price. Another People's Grocery project, Collards n' Commerce, is a sustainable agriculture and entrepreneurship program for West Oakland youth ages fifteen to eighteen. The young people help run the mobile market, work in the gardens, and take classes in cooking, nutrition, and business. Aside from getting food to people who need it, the teens learn what healthy food is and where it comes from. That's the idea behind community gardens that are sprouting in places as diverse as Cheyenne, Wyoming, and Huntsville, Alabama. "At first glance, you could dismiss the community garden," ecological artist Allen Green told the *East Bay Express* in Berkeley, California. "It can't grow enough food to matter; you could say it's an insignificant trifle. But it's in spaces like these that the experiment for another world will begin."

The San Francisco Bay area is home to another stealth weapon against obesity—the Edible Schoolyard. Since 1993, Berkeley's Martin Luther King Middle School has had an impressive two-acre organic garden created by Alice Waters, owner of the famed Chez Panisse restaurant. The school's largely low- and middle-income students learn how to plant and cultivate the garden, which supplies a cooking class at the school. There, kids learn how to prepare meals from fresh ingredients they've grown themselves. Roasting and eating a fresh-picked ear of corn is one of the students' first projects.

Now Waters is taking the concept one step further. In 2004, her Chez Panisse Foundation entered into a partnership with the Berkeley Unified School District to create a "school lunch initiative" district-wide that integrates organic gardening, cooking classes, and nutrition education into the school lunch program. For starters, Waters will help design and raise funds for a new dining commons at King Middle School serving fresh seasonal dishes with as much locally grown food as possible—her way of getting kids excited about healthy food. The hope is to replicate the dining commons at the rest of the city's schools. "Kids need to find these foods to be delicious so they fall in love with them, and it changes their lives forever," Waters told the *New York Times*.

In the beginning, some thought the Edible Schoolyard was an elitist concept, but the idea has caught on across the nation. Plots of land tended by students as young as kindergartners are up and running in 400 school districts in twenty-two states. Farmers' markets also were thought to be accessible only to the food-obsessed middle-class. But increasingly such markets, which sell fruits and vegetables often picked the same day, are coming to economically disadvantaged neighborhoods. For some years, the Women, Infants, and Children food assistance program has offered its clients vouchers to farmers' markets—and cooking classes on how to prepare seasonal produce. The National Farm to School Program is also gaining momentum, as schools increasingly opt to get their produce from local farmers.

People seem more eager than ever to change the way they think about food. The international Slow Food movement, dedicated to ecologically sound food production, home-cooked meals, and "living a slower, more harmonious rhythm of life" has more than 80,000 members worldwide. In 2004, some of the movement's leading lights gathered for a lecture at the University of California at Berkeley. After the 400-seat auditorium filled to capacity, hundreds of people—students, older and middle-aged people, young people in business attire—crowded into the lobby and down the front stairs, refusing to go home. The perplexed organizers had to agree to run a video of the event on the Internet. You would have thought Springsteen was performing.

The Ones in Charge

American kids live in both the wealthiest and the near-fattest nation on Earth. It is a nation built on determination and self-reliance, but kids can't rely on themselves to stop the epidemic of obesity. They don't need to start *making* better choices—they need to start *getting* better choices. They need to be able to select from a wide variety of foods that are easy to find, to afford, and to enjoy. They need a chance to walk to school and to play in the park and to have a real PE class taught by a real PE teacher.

Children, especially young children, can't be held accountable for what they eat and how much they exercise. Even in days gone by in environments where food was wholesome and healthy, children weren't allowed to choose their own food. Ultimately, despite all the distractions and temptations of modern life, adults are responsible for what they and their kids eat.

If we make just two changes in our complex lives to make things better for our kids—whether they're overweight or not—it should be bringing back the family meal and exercising with our kids once a day. Dinner at the table is really a gift for the entire family. After an exhaustive review of studies on the subject, researchers at Washington State University recently concluded that regular family meals improved language skills among preschoolers, boosted the grades of older kids, and, not surprisingly, increased everyone's consumption of healthier foods. By simply eating with their family, kids ate more vegetables, drank less soda, and took in fewer fatty and fried foods. In fact, whether or not a family had regular meals together turned out to be the most reliable predictor of how well-adjusted the children were in general.

Equally important, we should reexamine our own too-busy lives and find time just for active play with our children. That may mean that we stop checking our email and talking on the cell phone while keeping one eye cocked on the kids. Shooting hoops, riding bikes, or dancing to music with your children may seem like a small thing. But it frees them, however briefly, from the lure of television, video games, and the sedentary lifestyle that has helped pave the way for the obesity epidemic.

It's important to push the government to take a leadership role, but parents really can't afford to wait. Families must take action, whether that means joining a movement or joining each other for dinner. Change isn't easy no matter where it happens. But change is happening. We've seen it in families and communities from California to Connecticut. As for the rest of us, we can take to the basketball courts, or we can take to the streets. Either way, the exercise will do us good.

Notes

Chapter 1: An Epidemic for a New Age

Page

1 **Today, Dr. Cam-Tu Tran runs a small clinic:** From in-person interview with Cam-Tu Tran, conducted by Laurie Udesky, in February 2004.

2 **Seventeen hundred miles to the east, a Texas school administrator:** From in-person interview with Gina Castro, conducted by EH, May 2004.

2 **In just one school, she finds eight fourth graders:** Ibid.

3 **Americans now rank among the fattest people on Earth:** "Statistics Related to Overweight and Obesity," National Institute of Diabetes & Digestive & Kidney Diseases, Weight Control Information Network, July 2003.

3 **One third of U.S. adults are slightly to moderately overweight:** Ibid.

3 **caused some 400,000 adult deaths in 2000:** "Citing 'Dangerous Increase' in Deaths, HHS Launches New Strategies Against Overweight Epidemic," United States Department of Health, March 9, 2004.

3 **and obesity is poised to overtake tobacco:** Ibid.

3 **Sixteen percent of U.S. children:** Allison A. Hedley, et al., "Prevalence of Overweight and Obesity Among U.S. Children, Adolescents, and Adults, 1999–2002," *Journal of the American Medical Association*, June 16, 2004.

3–4 **The percentage of overweight youngsters ages 6 to 11 has tripled:** "Prevalence of Overweight Among Children and Adolescents: United States, 1999–2000," National Center for Health Statistics.

4 **"We're even seeing obesity in adolescents in India":** Emma Ross and Joseph Verrengia, "Obesity Spreads Worldwide," Associated Press, May 9, 2004.

4 **a disease that used to be found almost exclusively in adults:** "Study Will Identify Best Treatment for Type 2 Diabetes in Youth," U.S. Department of Health and Human Services, March 15, 2004.

4 **blood pressure is slowly creeping up:** "Consequences of Overweight in Children and Adolescents," Centers for Disease Control and Prevention, August 2002.

4 **the first clogging streaks of plaque:** Amanda Gardner, "Overweight Children at Higher Heart Risk," Healthfinder.gov, August 18, 2003.

4 ***Journal of the American Medical Association* (May 2004):** Paul Muntner, et al., "Trends in Blood Pressure Among Children and Adolescents," *Journal of the American Medical Association*, May 5, 2004.

4 **"I'm looking at nine- and ten-year-olds:"** From in-person interview with Barbara King Hooper, conducted by EH, January 2003."

7 **Dr. Fred Gunville witnessed this transformation firsthand:** From in-person interview with Fred Gunville, conducted by CW, May 2003.

7 **This disease occurs when the immune system attacks:** "Type 1 Diabetes," Endocrinology Health Guide, *University of Maryland Health Guide*, May 14, 2003.

7 **symptoms that often included fatigue, unquenchable thirst:** Ibid.

7 **Kids with a fasting reading of more than 100 milligrams of glucose:** "Insulin Resistance and Pre-Diabetes," National Diabetes Information Clearinghouse, May 2004.

8 **what was once only a trickle of kids with type 2:** From in-person interview with Fred Gunville, conducted by CW, May 2003.

8 **extra pounds can upset the body's delicate chemistry:** Gökhan Hotamisligil, "The Irresistible Biology of Resistin," *Journal of Clinical Investigation*, January 2003.

8 **a phenomenon that doctors call insulin resistance:** "Insulin Resistance and Pre-Diabetes," National Diabetes Information Clearinghouse, May 2004.

8 **creating a toxic sludge:** "What Is Type 2 Diabetes?" ReutersHealth, December 2001.

8 **but even that may not be enough:** "Insulin Resistance and Pre-Diabetes," National Diabetes Information Clearinghouse, May 2004.

8 **higher than normal–the hallmark of prediabetes:** Ibid.

8 **The insulin is not only much weaker than before:** "What Is Type 2 Diabetes?" ReutersHealth, December 2001.

8 **predicts that many children diagnosed with the disease:** From a lecture by William Dietz at the 2004 annual conference of the American Alliance for Health, Physical Education, Recreation, and Dance, New Orleans, March 2004.

9 **Dr. Suruchi Bhatia has had exactly the same experience:** From in-person interview with Suruchi Bhatia, conducted by EH, April 2003.

9 **the disease disproportionately affects the poor and the nonwhite:** *Healthy People 2010*, 2nd ed., U.S. Department of Health and Human Services.

10 **up to thirty-eight percent–as many as 21,000:** From interviews with former Oakland Unified School District nurse/diabetes educator, Joe Solowiejczyk, in February 2003, and Carole Flowers, Program Manager, Health Services, Oakland Unified School District, May 2004.

10 **fully one third of those born after 2000 will have type 2 diabetes:** Reported by the CDC at the American Diabetes Association's annual scientific sessions, June 2003.

10 **may be the first in history to have a shorter life expectancy:** Richard Carmona, Surgeon General, "The Growing Epidemic of Childhood Obesity," Department of Health and Human Services, congressional testimony, March 2, 2004.

10 **are likely to become severely overweight adults:** D.S. Freedman, et al., "Relationship of Childhood Obesity to Coronary Heart Disease Risk Factors in Adulthood: The Bogalusa Heart Study," *Pediatrics*, September 2001.

10 **may expect to live thirteen fewer years:** K. R. Fontaine, et al., "Years of Life Lost Due to Obesity," *Journal of the American Medical Association*, January 8, 2003.

10 **we're spending approximately $117 billion a year:** From a speech by U.S. Secretary of Health and Human Services Tommy Thompson, given at Time/ABC News Summit on Obesity, Williamsburg, Virginia, June 2, 2004.

10 **From 1979 to 1999, obesity tripled:** William Dietz, "Overweight in Childhood and Adolescence," *New England Journal of Medicine*, February 26, 2004.

10–11 **two thirds of parents of overweight children blamed themselves:** Vincent Iannelli, "Childhood Obesity: Who's to Blame?" pediatrics.about.com, 2004.

11 **eighty-seven percent said that individuals:** Time/ABC News poll, *Time Magazine*, June 7, 2004.

11 **treating obesity is "personal responsibility":** Speech by Health and Human Services Secretary Tommy Thompson, given at Time/ABC News Summit on Obesity, Williamsburg, Virginia, June 2, 2004.

11 **has led to a "toxic environment" that essentially overwhelms:** from *Food Fight*, Kelly D. Brownell and Katherine Battle Horgen, page 7.

11 **"Obesity has been considered a consequence of weak discipline":** Ibid., page 15.

11 **food industry produces 3,800 calories a day:** Marion Nestle, *Food Politics*, page 8.

11–12 **"There's something about human psychology":** Julie Creswell, "Chewing Out the Food Industry," *Fortune Magazine*, February 3, 2002.

12 **"We're becoming less responsible for our own health":** Radley Balko, "Beyond Personal Responsibility," Techcentralstation.com, May 17, 2004.

12 **"Especially in places like Texas":** From interview with Joan Miller, conducted by EH, May 2004.

12 **the fast-food industry alone spends $3 billion a year in advertising to children:** "The Role of Media in Childhood Obesity," The Henry J. Kaiser Family Foundation, February 2004.

12 **the federal government's $3.6 million yearly budget promoting consumption of fruits and vegetables:** Center for Science in the Public Interest online table showing 2001 advertising expenditure for federal 5 a Day Program.

13 **The highest rates of obesity occur among the poorest:** Adam Drewnowski, "Poverty and Obesity: the Role of Energy Density and Energy Costs," *American Journal of Clinical Nutrition*, January 2004.

13 **In 2002, thirty-five million Americans lived in households:** "State-by-State Hunger Fighting Trends Detailed in New FRAC Publication," Food Research and Action Center, April 21, 2004.

13 **twenty-six percent of adults with incomes below poverty:** Charlotte A. Schoenborn, et al., "Body Weight Status of Adults: United States, 1997–98," Centers for Disease Control and Prevention, September 6, 2002.

14 **In a study of low-income Latino families:** Patricia B. Crawford, et al., "How Can Californians Be Overweight and Hungry?" *California Agriculture*, January–March 2004.

14 **"Overweight has replaced malnutrition":** Statement by Patricia Crawford in *California Agriculture* press release, March 3, 2004.

14 **"If you're a poor person and you've got $5":** Marion Nestle in a panel discussion, "The Politics of Obesity, Confronting Our National Eating Disorder," UC Berkeley's Graduate School of Journalism, November 19, 2004.

14 **"The students don't do well in PE":** From in-person interview with Terre Logsdon, West Oakland, California, YMCA, conducted by EH, February 2004.

15 **both parents are working outside the home:** Tamar Lewin, "Two-Income Families Now the Norm, Census Bureau Says," *New York Times*, October 23, 2000.

15 **Americans work more than forty-nine hours a week:** Jill Andresky Fraser, *White-Collar Sweatshop*, page 20, citing the U.S. Bureau of Labor Statistics.

15 **"means twice as much potential for overwork":** Ibid., page 23.

15 **"I have enough resources financially":** From a talk by Joan B. Carter, Baylor College of Medicine, at the Casey Journalism Center on Children and Families, March 15, 2004.

16 **have to overcome at least one "serious physical hazard":** Bruce Appleyard, "Safe Routes to School: Introduction," National Center for Bicycling and Walking.

16 **"You know, I read about people in France":** From interview with Suruchi Bhatia, conducted by EH, April 2003.

16 **reported in 2002 not having enough textbooks:** "NEA/AAP Survey Finds Nationwide Textbook Shortages," National Education Association news release, October 8, 2002.

17 **"When we drop off kids at school":** From a phone interview with Margo Wootan, conducted by EH, June 2004.

17 **high schools provide daily phys ed:** "Healthy Schools for Healthy Kids," Robert Wood Johnson Foundation report, 2003.

17 **is costing schools millions of dollars each year:** "Students Poor Nutrition and Inactivity Comes with Heavy Academic and Financial Costs to Schools," press release from Action for Healthy Kids, September 23, 2004.

18 **overweight kids have a worse quality of life:** Charnicia E. Huggins, "Obese Kids Have Quality of Life on Par with Cancer," Reuters, April 8, 2003.

19 **"They call me fat":** From an in-person interview with Yadira Renteria, conducted by Psyche Pascual, May 25, 2004.

19 **"She was eating salads, walking, but her weight":** From an in-person interview with Patricia Orozco, conducted by PP, May 25, 2004.

19 **Researchers from Yale University showed:** J. D. Latner and A. J. Stunkard, "Getting Worse: The Stigmatization of Obese Children," *Obesity Research*, March 2003.

20 **kids' quality of life was far inferior:** Charnicia E. Huggins, "Obese Kids Have Quality of Life on Par with Cancer," Reuters, April 8, 2003.

20 **fifty-nine percent of the kids said they have tried:** "New Survey Reveals Kids Worried About Obesity, Too," news release from KidsHealth.org, January 13, 2004.

20 **often their parents who are in denial:** Amy E. Baughcum, "Maternal Perceptions of Overweight Preschool Children," *Pediatrics*, December 2000.

20 **"It's such a long progressive thing":** From an in-person interview with Melanie Ritsema, conducted by EH, May 2004.

21 **who calls the obesity epidemic "the terror within":** "Surgeon General to Cops: Put Down the Donuts," CNN.com, March 2, 2003.

21 **"We need to lead a cultural transformation":** Kim Severson, "Obesity 'A Threat' to U.S. Security: Surgeon General Urges Cultural Shift," *San Francisco Chronicle*, January 7, 2003.

21 **weight and height into account:** "BMI: Body Mass Index," Centers for Disease Control, April 17, 2003.

21 **over thirty makes them officially:** Ibid.

21 **above the ninety-fifth percentile on the growth charts:** "BMI for Children and Teens," Centers for Disease Control and Prevention, April 8, 2003.

Chapter 2: Fat City

Page

23 **by another, less festive name: Fat City:** From Scott Huddleston, "Image of S.A. as Walkers' Dream Staggers Some," *San Antonio Express-News*, March 11, 2004, and from interviews, conducted by EH in May 2004 with Joan Miller.

23 **San Antonio is the fattest city in America:** Michael Precker, "Taking-up-Space City," *Dallas Morning News*, October 3, 2003, citing a 2003 Centers for Disease Control and Prevention survey.

23 **study from *Men's Fitness* magazine:** "America's Fattest Cities 2004," *Men's Fitness* magazine, www.mensfitness.com/rankings/200.

24 **sixty-five percent of the city's adults are overweight:** 2002 Community Health Assessment and Health Profiles, Bexar County Community Health Collaborative, San Antonio Metropolitan Health District.

24 **from twenty-five to thirty-one percent are obese:** Figures from Ibid. and the CDC.

24 **In some neighborhoods, as many as seventy-six percent:** 2002 Community Health Assessment and Health Profiles, Bexar County Community Health Collaborative, San Antonio Metropolitan Health District.

24 **In San Antonio, partying and recreational eating:** From numerous in-person interviews in San Antonio; also Edie Jarolim, *Frommer's San Antonio & Austin*.

24 **"Ask anyone in town, 'Where's the buffet?'":** From in-person interview with Anna Guerrero, conducted by EH, May 2004.

24 **San Antonio has mobilized every key player:** From an in-person interview with Joan Miller, conducted by EH, May 2004.

24 **"a model for the rest of the nation":** *The Collaborative* newsletter, Spring 2003.

25 **got its first wake-up call in 1998:** From in-person interview with Joan Miller, conducted by EH, May 2004.

25 **nearly half the county's residents were sedentary:** 1998 Bexar County Community Health Assessment Report Summary.

25 **"It was too critical to avoid any longer":** From in-person interview with Joan Miller, conducted by EH, May 2004.

25 **he's lowered dangerously high blood sugars:** Oralia Garcia, et al., "Bienestar Health Program: A Comprehensive Approach to Reversing Hyperglycemia in Low-Income Children," *Diabetes*, June 2003.

26 **"Everybody knows everybody":** From in-person interview with Joan Miller, conducted by EH, May 2004.

26 **San Antonio's 417 square miles:** San Antonio Economic Development Foundation.

26 **"Community–That's what it's all about":** Stoneridge Homeowners Association website, http://www.stoneridgehoa.org.

27 **leading residents to use the term "09ers":** Edie Jarolim, *Frommer's San Antonio & Austin*.

27 **annual $3.51 billion from tourism:** San Antonio Economic Development Foundation.

27 **highest proportion of Hispanic residents:** 2002 Community Health Assessment and Health Profiles, Bexar County Community Health Collaborative, San Antonio Metropolitan Health District.

27 **its people have the lowest incomes:** Ibid.

27–28 **more than seventy-five percent of this area of San Antonio's adult residents were overweight:** Ibid.

28 **forty percent were obese:** Ibid.

28 **Of the 11,000 children in the WIC program, twenty percent:** From in-person interviews with Anna Guerrero and Melanie Ritsema at WIC, conducted by EH, May 2004.

28 **One family in particular stands out:** From in-person interview with Anna Guerrero, conducted by EH, May 2004.

28 **"We don't want to punish ourselves":** Ibid.

29 **"They aren't thinking five years down the road":** Ibid.

29 **say they face overwhelming obstacles:** From in-person interview with Melanie Ritsema and Anna Kraft, conducted by EH, May 2004.

29 **"Poverty kind of limits you":** From in-person interview with Melanie Ritsema, conducted by EH, May 2004.

29 **only twenty-seven percent believed their child's weight was a significant health risk:** Carolina Luna-Pinto, et al., "Hispanic Mothers' Perceptions of Obesity in Their Children," abstract presented to American Public Health Association annual meeting, October 22, 2001.

30 **diabetes is now the fourth-leading cause of death:** 2002 Community Health Assessment and Health Profiles, Bexar County Community Health Collaborative, San Antonio Metropolitan Health District.

30 **Eleven percent of the city's adults have the disease:** Ibid.

30 **Of 1,420 elementary school children:** From in-person interview with Roberto Treviño, conducted by EH, May 2004.

31 **"She was the only breadwinner":** From in-person interview with Melanie Ritsema, conducted by EH, May 2004.

31 **"don't get the message until you mention 'fat'":** Ibid.

32 **doctors rarely address the issue of obesity in children:** "Childhood Obesity Often Missed," *Health Scout News*, May 29, 2003.

32 **they didn't feel they had the skills to counsel kids effectively:** Mary T. Story, et al., "Management of Child and Adolescent Obesity: Attitudes, Barriers, Skills, and Training Needs Among Health Care Professionals," *Pediatrics*, July 2002.

32 **"Just the basics":** From in-person interview with Anna Guerrero, conducted by EH, May 2004.

32 **paying out of pocket for the sugar-coated kind:** From in-person interview with Anna Kraft, conducted by EH, May 2004.

32 **WIC sites have vending machines operated by the city:** From in-person interview with Melanie Ritsema, conducted by EH, May 2004.

33 **although vending machines aren't a critical element:** From in-person interview with Joan Miller, conducted by EH, May 2004.

33 **"They're not going anywhere":** From in-person interview with Melanie Ritsema, conducted by EH, May 2004.

33 **In the spring of 2004, McDonald's Corporation unveiled:** "McDonald's Brings 'Go Active! American Challenge' to San Antonio with Olympic Athlete Josh Davis & Ronald McDonald," McDonald's press release, May 18, 2004.

33 **at the site of Texas's most famous massacre:** Ibid.

34 **Walk San Antonio, an ambitious program that's enrolled more than 10,000:** From a speech by Joan Miller, May 18, 2004.

34 **and Miller's got the scars to prove it:** From in-person interview with Joan Miller, conducted by EH, May 2004.

34 **"There are now more overweight people in the world":** From a speech by Melanie Ritsema, May 18, 2004.

35 **to consider BMI as a fifth vital sign:** Ibid.

35 **"We hope this will inspire all restaurants":** Ibid.

35 **"There's nothing like having that medal draped":** From a speech by Josh Davis, May 18, 2004.

35 **"I'm so happy McDonald's is providing us":** Ibid.

35 **"Kids love it, families love it":** Ibid.

35 **enjoying an order of hash browns and a stack:** From an in-person interview with Jim Hinkle at Bellaire Elementary School, conducted by EH, May 2004.

35–36 **His mother brought the meal to school as a treat:** Ibid.

36 **already is showing signs of acanthosis nigricans:** From a visit to Bellaire Elementary School by EH, May 2004.

36 **"The people over there, they don't come to our side":** From in-person interview with Gina Castro, conducted by EH, May 2004.

36 **Her husband died seven years ago at age forty-nine:** Ibid.

36 **has already lost a leg to the disease:** Ibid.

36 **"He knows how his father died":** Ibid.

37 **a Texas-based elementary school program:** "Healthy Schools for Healthy Kids," a report of the Robert Wood Johnson Foundation, 2003.

37 **program is being used in thirty U.S. states:** Ibid.

37 **the CATCH program significantly increased moderate to vigorous:** "School Program Reduces Risk of Obesity, Disease in Border Community," news release, Center for the Advancement of Health, July 24, 2002.

37 **"I'm the new food police":** From in-person interview with Gina Castro, conducted by EH, May 2004.

37 **found thirty-eight kids with blood sugars high enough:** Ibid.

37 **number of children with high blood pressure:** Ibid.

37 **"Our nurses are so overworked":** Ibid.

37 **"My biggest problem," she hesitates to actually say it:** Ibid.

37 **"She came to school wearing a women's size extra large":** Ibid.

38 **"children are coming in very obese in kindergarten":** Ibid.

38–39 **Some parents are appalled:** Ibid.

39 **accused the schools of meddling and harming:** Michelle Galley, "School Letters on Students' Obesity Outrage Some Parents," *Education Week*, April 3, 2002.

39 **even appeared on the *Today* show to complain:** Ibid.

39 **Coach Jim Hinkle is warming up the kids:** From a visit to Bellaire Elementary School in San Antonio by EH, May 2004.

39 **forty-five minutes of phys ed four days a week:** From in-person interview with Gina Castro, conducted by EH, May 2004.

40 **"Simon says how many of you know about the food groups?":** From a visit to Bellaire Elementary School in San Antonio by EH, May 2004.

40 **"I worked the hell out of them":** From in-person interview with Jim Hinkle, conducted by EH, May 2004.

40 **this generation will find exercise they like:** Ibid.

40 **Hinkle's students successfully completed a yearlong marathon:** Ibid.

41 **"We don't talk to the parents about that":** Ibid.

41 **Hinkle sells the kids *paletas*, Spanish for Popsicle:** Ibid.

42 **decided last year to ban sweet and fatty treats:** From in-person interview with Kaye Lucas, conducted by EH, May 2004.

42 **her kids are calmer in class:** Ibid.

42 **"I can't open this":** From a visit to Bellaire Elementary School in San Antonio, conducted by EH, May 2004.

42 **state of Texas for guidance, where fatty foods were cut way back:** Texas Public School Nutrition Policy, effective August 1, 2004; http://www.agr.state.tx.us/foodnutrition/policy/food_nutrition_policy.pdf.

42 **junk-food sales have taken a nosedive:** From in-person interview with Sally Cody, conducted by EH, May 2004.

42 **Her budget has taken a hit for it too:** Ibid.

42 **Take Flaming Hot Cheetos, for instance:** Ibid. Just about everywhere we traveled, Flaming Hot Cheetos came up in conversation, either as kids' preferred snack food or as a metaphor for how badly kids are eating.

42 **"They're really good":** From in-person interview with Sally Cody, conducted by EH, May 2004.

43 **a four-fat-gram version designed specially:** Ibid.

43 **"I don't even want to know how much fat":** Ibid.

43 **It means she loses more than $400,000 a year:** Ibid.

43 **used to send her "mountains" of shortening and butter:** Ibid.

43 **Wednesday is enchilada day:** Ibid.

44 **"They've made it taste wonderful":** Ibid.

44 **Today, there's not a single deep-fat fryer left:** From in-person interview with Gina Castro, conducted by EH, May 2004.

44 **children used to eat better when she started twenty-five years ago:** From in person interview with Carmen Martinez, conducted by EH, May 2004.

44 **used to be a hoodlum, running the streets:** From in-person interview with Roberto Treviño, conducted by EH, May 2004.

45 **the advanced type 2 diabetes, rampant in his community:** Ibid.

45 **"You go around the country and all you hear":** Ibid.

45 **he'd have to start with children:** Ibid.

45 **a handsome mustached man astride a white horse:** From a visit to Bienestar by EH, May 2004.

46 **"This is a mission":** From in-person interview with Roberto Treviño, conducted by EH, May 2004.

46 **"If you go to work and your coworkers are telling you":** Ibid.

46 **In 1999, with the help of a $2 million grant:** Ibid.

46 **Over each of the thirty-two-week trials, blood sugar decreased:** Oralia Garcia, et al., "Bienestar Health Program: A Comprehensive Approach to Reversing Hyperglycemia in Low-Income Children," *Diabetes*, June 2003.

47 **Body fat decreased but not significantly:** From in-person interview with Roberto Treviño, conducted by EH, May 2004.

47 **at age nine it's almost too late for that:** Ibid.

47 **"When we do kindergarten through middle school":** Ibid.

47 **about $190 a year, compared to $1,200 a year:** Statistics compiled by Bienestar.

47 **Alma Lopez has been on a healthier eating quest:** From in-person interview with Alma Lopez and Ameri Lopez, conducted by EH, May 2004.

48 **"I'm Mexican, and you know you cook with a lot of oil":** Ibid.

48 **has a long family history of diabetes:** Ibid.

48 **"I like when I learn that I shouldn't watch TV too much":** Ibid.

48 **"Now I like to play kickball, and I also like to play soccer":** Ibid.

48 **"One of my friends will even lick the bag":** Ibid.

49 **had Ameri's blood sugar tested:** Ibid.

49 **"I look out the window and wonder":** From in-person interview with Roberto Treviño, conducted by EH, May 2004.

Chapter 3: Babes in Calorie Land

Page

51 **now open in forty-nine states (sorry, Hawaii):** Golden Corral Franchise Information.

51 **pour gummy bears on fudge brownies:** CW visited the Golden Corral in Billings, Montana, June 2004.

52 **offers the following dinner menu for the preschool child:** Ruth Berolzheimer, *Culinary Arts Institute Encyclopedic Cookbook* (Culinary Arts Institute, 1966).

52 **summed up the situation at the 2004 Summit on Obesity in Williamsburg, Virginia:** Webcast of presentation available at Robert Wood Johnson Foundation website: http://www.rwjf.org/news/index.jhtml. (Click on "webcasts.")

52 **fast food cartons, snack bags, and candy wrappers:** From Samara Nielsen, et al., "Trends in Energy Intake in the U.S. between 1977 and 1996: Similar Shifts Seen Across Age Groups," *Obesity Research*, May 2002. The study noted a sharp increase in consumption of french fries, pizza, salty snacks, and candy.

52 **A hundred years ago, many children suffered from rickets because of a shortage vitamin D:** From a lecture by William Dietz at the 2004 annual conference of the American Alliance for Health, Physical Education, Recreation, and Dance, New Orleans, March 2004.

52 **Others developed cretinism (a form of mental retardation) because they weren't getting enough iodine:** Ibid.

53 **Malnourishment left them spindly, sickly, and short:** Ibid.

53 **the nutritional disease of a new age:** Ibid.

53 **started picking up steam in the mid-1970s:** Cynthia Ogden, et al., "Epidemioloical Trends in Overweight and Obesity," *Endocrinology and Metabolism Clinics of North America*, vol. 32, 2003.

53 **salty snacks more than doubled in those nineteen years:** "Trends in Energy Intake in the U.S. between 1977 and 1996: Similar Shifts Seen Across Age Groups," *Obesity Research*, May 2002.

53 **soda also rose sharply across the board:** Ibid.

53 **eighty-three calories' worth of sweeteners a day to their diet (not including sugar naturally found in foods) during those years:** Barry Popkin and Samara Nielsen, "The Sweetening of the World's Diet," *Obesity Research*, November 2003.

53 **accounted for fifty-four of those extra calories:** Ibid.

53 **decline slightly between the early 1970s and 2000:** Ronette Briefel and Clifford Johnson, "Secular Trends in Dietary Intake in the United States," *Annual Review of Nutrition*, February 11, 2004.

53 **thirty-three percent (or one third) of calories from fat in 2000:** Ibid.

53 **the recommended limit of thirty percent:** Ibid.

53 **largely thanks to breakfast cereal, corn chips, pretzels, popcorn, and crackers:** Ibid.

53 **less than one fourth of people got the recommended daily servings:** Ibid.

53–54 **which range from six a day for young children to eleven for active men:** "Dietary Guidelines: Build a Healthy Base," U.S. Department of Health and Human Services.

54 **more fruits and vegetables in the 1990s:** Ronette Briefel and Clifford Johnson, "Secular Trends in Dietary Intake in the United States," *Annual Review of Nutrition*, February 11, 2004.

54 **thirty percent of vegetables were potatoes:** Ibid.

54 **one fourth of their calories from desserts and sweeteners:** Ashima Kant, "Reported Consumption of Low-Nutrient-Density Foods by American Children and Adolescents," *Archives of Pediatrics and Adolescent Medicine*, August 2003.

54 **and by seventy-four percent in adolescent boys:** Marie-Pierre St-Onge, et al., "Changes in Childhood Food Consumption Patterns: A Cause for Concern in Light of Increasing Body Weights," *American Journal of Clinical Nutrition*, 2003.

54 **Boys now average about nineteen ounces of soda a day:** David Ludwig, et al., "Relation Between Consumption of Sugar-Sweetened Drinks and Childhood Obesity: A Prospective, Observational Analysis," *The Lancet*, February 17, 2001.

54 **cut back on milk by thirty-six percent between the 1960s and the 1990s:** C. Cavadini, et al., "US Adolescent Food Intake Trends from 1965 to 1996," *Western Journal of Medicine*, December 2000.

54 **meet all of the recommendations of the USDA food pyramid for children:** Marie-Pierre St-Onge, et al., "Changes in Childhood Food Consumption Patterns: A Cause for Concern in Light of Increasing Body Weights," *American Journal of Clinical Nutrition*, 2003.

54 **those chips, buns, fries, and sodas have something in common:** Arnold Slyper, "The Pediatric Obesity Epidemic: Causes and Controversies," *The Journal of Clinical Endocrinology & Metabolism*, June 2004.

55 **last thing on Tracy Graham's mind as she fed her family:** From phone interview with Tracy Graham, conducted by CW, May 2004.

55 **prepared outside of the house, a near 100 percent increase since 1977:** Joanne Guthrie, et al., "Role of Food Prepared Away from Home in the American Diet, 1977–78 Versus 1994–1996: Changes and Consequences," *Journal of Nutrition Education and Behavior*, May/June 2002.

56 **on snacks and meals prepared away from home:** Ibid.

56 **"Food away from home has become an affordable treat":** From phone interview with Joanne Guthrie, conducted by CW, April 2004.

57 **fivefold between the late-1970s and the mid-1990s:** Shanthy Bowman, et al., "Effects of Fast-Food Consumption on Energy Intake and Diet Quality Among Children in a National Household Survey," *Pediatrics*, January 1, 2004.

57 **about thirty percent of kids will eat fast food, and many eat it every day:** Ibid.

57 **250,000 fast food restaurants in this country:** Ibid.

57 **with two pounds of beef:** A Whopper usually sells for about $2.40. At typical grocery store prices, that's enough for two pounds of ground beef, a bun, and a few condiments.

57 **costs a little over $9:** Price at a KFC in Billings, Montana, in June 2004.

57 **costs a little over $4:** Diana Sagers, et al., "The Cost of Convenience," Utah State University Extension Office.

57 **you don't have to look further than the produce aisle:** Lee Reich, "USDA: Fruits and Veggies Not as Pricey as You Think," *USA Today*, August 1, 2004.

58 **it can all be yours:** Cited on Calorieking.com.

58 **for about $2.30:** Price at a Billings, Montana, supermarket in June 2004.

58 **cost less than the calories in fruits, vegetables, meat, and other unprocessed foods:** Judith Blake, "Researcher Links Food Prices, Obesity," *St. Paul Pioneer Press*, October 17, 2003.

59 **children's menus of most major chain restaurants are a nutritional disaster:** Jayne Hurley and Bonnie Liebman, "Kid's Cuisine: What Would You Like With Your Fries?" *Nutrition Action Healthletter*, March 2004.

59 **900 calories and more saturated and trans fats than a kid should eat in an entire day:** Nutritional Facts, Ibid.

60 **920 calories and forty-two grams of fat in one meal:** Nutritional Facts, Burger King website.

60 **187 extra calories on days when they eat fast food:** Shanthy Bowman, et al., "Effects of Fast-Food Consumption on Energy Intake and Diet Quality Among Children in a National Household Survey," *Pediatrics*, January 1, 2004.

60 **"equivalent to less than a cup of lima beans . . . Horrors!":** "Fast Food Study Over-Hyped," Center for Consumer Freedom, January 5, 2004.

60 **"about 3,400 extra calories to gain a pound":** From phone interview with Shanthy Bowman, conducted by CW, March 2004.

60 **concluded overweight children have "some sort of susceptibility" to fast food:** From phone interview with Cara Ebbeling, conducted by CW, June 2004. See also Michael Lasalandra, "Fast Food Plus Fat Kids May Equal Supersized Problem," *Boston Herald*, October 16, 2003.

61 **also likely to overeat when given outrageously large portions:** From phone interview with Jennifer Fisher, conducted by CW, April 2004.

61 **kids ate twenty-five percent more than they did before:** Alfredo Flores, "Larger Portions May Lead Children to Overeat," Agricultural Research Service, July 16, 2003.

61 **all increased dramatically between 1977 and 1998:** Ronette Briefel and Clifford Johnson, "Secular Trends in Dietary Intake in the United States," *Annual Review of Nutrition*, February 11, 2004.

61 **a jump that added 136 extra calories:** Ibid.

61 **increasingly large servings milk, bread, cereal, juices, and peanut butter:** Ibid.

61 **"huge platters of food coming out of the kitchen":** From phone interview with Melanie Polk, conducted by CW, November 2001.

62 **about 250 calories and seventeen teaspoons of sugar:** Sources differ on how much sugar can be found in sodas, partly because the amount varies from one brand to another. In "Sugar Consumption 'Off the Charts,' Say Health Experts," the Center for Science and the Public Interest reports that a typical twelve-ounce soft drink contains nine teaspoons of sugar, which works out to fifteen teaspoons in a twenty-ounce serving. A report from the Colorado State University Extension Office–"Are We Drinking Too Much Soda Pop?"–says each twelve-ounce serving of soda contains ten to twelve teaspoons of sugar.

62 **less expensive and more attractive to young consumers:** George Bray, et al., "Consumption of High-Fructose Corn Syrup in Beverages May Play a Role in the Epidemic of Obesity," *American Journal of Clinical Nutrition*, April 2004.

62 **"much sweeter and gives more of a bang for the buck":** From email interview with George Bray, conducted by CW, June 2004.

62 **corn syrup doesn't trigger the release of leptin, a hormone that tells the body when it's full:** George Bray, et al., "Consumption of High-Fructose Corn Syrup in Beverages May Play a Role in the Epidemic of Obesity," *American Journal of Clinical Nutrition*, 2004.

62 **couldn't understand why her daughter weighed more than 300 pounds:** From phone interview with Melinda Sothern, conducted by CW, October 2001.

63 **raise the risk of obesity by sixty percent:** David Ludwig, et al., "Relation Between Consumption of Sugar-Sweetened Drinks and Childhood Obesity: A Prospective, Observational Analysis," *The Lancet*, February 17, 2001.

63 **published a fact sheet–"Straight Facts about Beverage Choices":** Published in the *Journal of the American Dietetic Association*, September 2001.

63 **$25,000 for the soda "fact sheet":** From phone interview with Lori Ferme, media relations manager for the American Dietetic Association, conducted by CW, January 2003.

63 **impression that the dietetic association endorses soda consumption:** By policy of the American Dietetic Association, all "fact sheets" expire within three years of publication (some sooner). And once a fact sheet expires, the sponsor can no longer display it on its website. As of July 2004, "Straight Facts About Beverage Choices" was no longer available on the National Soft Drink Association's website. For more on the fact sheet policy of the ADA, see "Nutrition Fact Sheet Policy," American Dietetic Association.

64 **"ADA has final editorial control of the fact sheet content":** "Nutrition Fact Sheet Policy," American Dietetic Association.

64 **didn't prove that soda actually *caused* the weight gain:** "Food Police Do a Body Bad," Center for Consumer Freedom (at CSPIscam.com), April 21, 2004.

64 **found no link between soda and obesity in children ages twelve to nineteen:** R. A. Forshee and M. L. Storey, "Total Beverage Consumption and Beverage Choices Among Children and Adolescents," *International Journal of Food Science and Nutrition*, July 2003.

64 **grant from, naturally, the National Soft Drink Association:** Ibid. See also Robert Preidt, "What's Behind Teen Beverage Choices," *USA Today*, June 23, 2003.

64 **compared children's weights with their soda habits:** R. A. Forshee and M. L. Storey, "Total Beverage Consumption and Beverage Choices Among Children and Adolescents," *International Journal of Food Science and Nutrition*, July 2003.

64 **can't come close to proving–or disproving–cause and effect:** This hasn't stopped supporters of soda from drawing some bizarre conclusions. In June 2003, Sarah Theodore, the editor of *Beverage Industry*, noted that the study found a "higher rate of overweight kids who drink diet soft drinks." It's hardly a shocking discovery. How many kids would subject themselves to the taste of a diet soda if they weren't already concerned about their weight? Theodore, however, says the finding is "very interesting" because it suggests that "weight problems cannot be attributed simply to beverage consumption." In other words, because heavy kids drink diet soda, regular sugary sodas can't be contributing to obesity.

64–65 **an incredibly simple–and incredibly effective–effort to curb childhood obesity:** Janet James, et al., "Preventing Childhood Obesity by Reducing Consumption of Carbonated Drinks: Cluster Randomized Control Trial," *British Medical Journal* (online edition), April 27, 2004.

65 **proposal for a two-cent-per-can "soda tax" to fund school health programs in 2002:** "California Soft Drink Proposal Fails," *Food and Drink Weekly*, May 6, 2002.

65 "**understand what the real problem is":** Quoted from Tom Bachmann, "Why Are Our Kids Getting Fat?" *Beverage Industry*, May 2002.

65 **heavy lobbying from Big Soda:** "Tax Soda, Help Kids' Obesity," CBSNews.com, March 28, 2002.

65 **the proposed tax was defeated:** "California Soft Drink Proposal Fails," *Food and Drink Weekly*, May 6, 2002.

65 **heavy-handed letter to Marion Nestle:** The letter, dated March 27, 2002, was sent by Venable Attorneys at Law at the behest of the Sugar Association.

65–66 **Nestle explained what *really* is commonly known by experts in the field of nutrition:** Nestle's letter, dated April 5, 2002, is addressed to Jeffrey S. Tenenbaum of Venable Attorneys at Law.

66 **has more fat and calories than a Big Mac:** "Bunless Burger vs. The Fluffernutter," Salon.com, May 5, 2004.

66 **more fat than two Snickers bars:** Ibid.

66 **more calories from fat than a Quarter Pounder:** McDonald's Nutrition Information.

66 **gained twenty-four pounds by eating at McDonald's three times a day for a month:** Jody Genessy, "Fast-Food Flick Is a Supersized Hit," *Deseret Morning News*, January 21, 2004.

66 **McDonald's started phasing out its "supersize" option:** "McSupersizes to be Phased Out," CNN.Com, March 3, 2004.

66 **was not a response to the movie:** For example, see Ibid.

66 **plans to release a children's exercise video featuring Ronald McDonald in 2005:** "Ronald McDonald Romps in Exercise Video," Reuters, June 8, 2004.

66 **passed the bill by a wide margin in March 2004:** "House Passes Cheeseburger Bill," CBSNews.com, March 10, 2004.

67 **but none was successful:** For an in-depth discussion of the unsuccessful lawsuit filed by the two teenagers, see Sherry Kolb, "Why Suing McDonald's Could be a Good Thing," CNN.com, January 29, 2003.

67 **can't be held responsible for the poor choices of their customers:** Emily Heller, "Fat Suit Weighs In," *The National Law Journal*, December 11, 2002.

67 **"it's not our bad food, it's people's lack of responsibility":** From phone interview with Michael Lowe, conducted by CW, February 2004.

67 **no *proof* that any of these foods contribute to childhood obesity:** "Fast Food Study Over-Hyped," Center for Consumer Freedom, January 5, 2004. See also "Food Police Do a Body Bad," Center for Consumer Freedom (at CSPIscam.com), April 21, 2004.

67 **and they're right:** From phone interview with Cara Ebbeling, conducted by CW, June 2004.

67 **recruit thousands of children for an unprecedented study:** The approach described here is a randomized, controlled, and longitudinal study, the gold standard for this type of research. (Longitudinal means subjects are followed over time.) Such studies are notoriously difficult to conduct, especially when they involve human behaviors or lifestyles.

67 **don't eat a lot less at their next meal:** France Bellisle and Marie-Francoise Rolland-Cachera, "How Sugar-Containing Drinks Might Increase Adiposity in Children," *The Lancet*, February 17, 2001. Researchers sometimes refer to the calories in sugary drinks as "stealth" calories because they don't make a person feel full. As this article explains, there's no evidence that people compensate for those extra calories by cutting back on other foods and beverages.

67–68 **walking for two and a half miles, which is what it would take to burn those extra soda calories:** James Sallis has estimated that a typical child burns fifty calories on a half-mile walk, so it would take 2.5 miles to burn off the 250 calories in a twenty-ounce bottle of Coke.

68 **"What we are actually seeing is that child obesity is increasing":** From a phone interview with Shanthy Bowman, conducted by CW, February 2004.

68 **"if it's more palatable, eat more of it":** From a phone interview with Michael Lowe, conducted by CW, February 2004

68 **have a remarkable ability to regulate their calorie intake:** Ibid.

68 **was practically dogma among child psychologists in the 1960s:** From in-person interview with William Woolston, child psychologist, conducted by CW, April 2004.

68 **often called "the wisdom of the body":** Leann Birch, "Development of Food Preferences," *Annual Review of Nutrition*, vol. 19, 1999.

69 **many complex factors, including the hardwiring in our brains and the food we eat early in our childhood:** Ibid.

69 **and a reluctance to try anything new:** Ibid.

69 **even soothe a baby boy during circumcision!:** Ibid.

69 **kids very quickly learn to prefer foods packed with calories:** Ibid.

69 **children needed every calorie they could get in order to grow:** Ibid. Also from phone interview with Michael Lowe, conducted by CW, March 2004.

69 **"food is omnipresent, inexpensive, and very palatable":** From phone interview with Michael Lowe, conducted by CW, March 2004.

70 **$12 billion a year on ads aimed at kids:** "Television Advertising Leads to Unhealthy Habits in Children, Says APA Task Force," American Psychological Association, February 23, 2004.

70 **$3 billion of which comes from fast food companies:** "The Role of Media in Childhood Obesity," the Henry J. Kaiser Family Foundation, February 2004.

70 **$3.6 million each year on its Five-A-Day campaign that promotes fruits and vegetables:** Neal Barnard, "The War Against the Messenger; Trashing the Food Police," Physicians Committee for Responsible Medicine, May 2002.

70 **watches 40,000 commercials in a year, twice as many as in the late 1970s:** "The Role of Media in Childhood Obesity," the Henry J. Kaiser Family Foundation, February 2004.

70 **nine percent are for fast food:** Ibid.

70 **eleven food commercials per hour of television:** Ibid.

70 **children under eight generally don't understand the purpose of advertisements:** "Television Advertising Leads to Unhealthy Habits in Children, Says APA Task Force," American Psychological Association, February 23, 2004.

71 **the kids who watched the most TV were most likely to make the wrong choice:** "The Role of Media in Childhood Obesity," the Henry J. Kaiser Family Foundation, February 2004.

71 **child-focused marketing goes far beyond television:** From phone interview with Cara Ebbeling, conducted by CW, June 2004.

71 **can play dozens of edifying games that all feature Nabisco products:** For an in-depth review of online advertising games, see Joseph Pereira, "Online Arcades Draw Fire for Immersing Kids in Ads," *Wall Street Journal*, May 3, 2004.

72 **Moving to a new school is never easy:** From phone interview with Clay Jones, conducted by CW, May 2003.

73 **kids simply can't "learn" to eat healthier food:** From phone interview with Michael Lowe, conducted by CW, March 2004. See also Michael Lowe, "Self-Regulation of Energy Intake in the Prevention and Treatment of Obesity: Is It Feasible?" *Obesity Research*, 2003.

73 **gaining more weight than the kids who didn't diet:** Alison Field, et al., "Relationship Between Dieting and Weight Change Among Preadolescents and Adolescents," *Pediatrics*, October 2003.

73 **"akin to asking if taking drugs is desirable":** Michael Lowe, "Self-Regulation of Energy Intake in the Prevention and Treatment of Obesity: Is It Feasible?" *Obesity Research*, October 2003.

74 **kidney stones:** F. Lefevre and N. Aronson, "Ketogenic Diet for the Treatment of Refractory Epilepsy in Children: A Systematic Review of Efficacy," *Pediatrics*, April 2000.

74 **calcium loss in bones:** Stephen Rothman, "Ketogenic Diet Lacks Vitamin and Mineral Support," *Medical Post*, January 18, 2000.

74 **high cholesterol:** P. Kwiterovich, et al., "Effect of a High-Fat Ketogenic Diet on Plasma Levels of Lipids, Lipoproteins, and Apolipoproteins in Children," *Journal of the American Medical Association*, August 20, 2003.

74 **and heart trouble:** "Study Warns of Potential Dangers of a Ketogenic Diet," Cincinnati Children's Hospital Medical Center, July 6, 2000.

74 **most people eventually return to a normal diet and regain all or most of the weight:** See, for example, G. D. Foster, et al., "A Randomized Trial of a Low Carbohydrate Diet for Obesity," *New England Journal of Medicine*, May 22, 2003. This study found that subjects on a low-carbohydrate diet quickly lost weight and then quickly gained it back. After one year, they hadn't lost any more weight than people on a low-fat diet.

74 **a small study of "reduced-glycemic load" diets on sixteen overweight adolescents ages thirteen to twenty-one:** Cara Ebbeling, et al., "A Reduced-Glycemic Load Diet in the Treatment of Adolescent Obesity," *Archives of Pediatrics and Adolescent Medicine*, August 2003.

74 **"couple of different ways to reduce glycemic load":** From phone interview with Cara Ebbeling, conducted by CW, June 2004.

74 **other kids who were put on a traditional low-fat diet:** Ibid.

75 **requiring restaurants to clearly list the caloric content of every item on the menu:** Michael Lowe, "Self-Regulation of Energy Intake in the Prevention and Treatment of Obesity: Is It Feasible?" *Obesity Research*, 2003.

Chapter 4: Schools–From Sellouts to Sanctuaries

Page

77 **she didn't anticipate a full-scale nutritional nightmare:** From an in-person interview with Linal Ishibashi, conducted by EH, January 2004.

77 **were making a lunch of a bottle of soda and an extra-large bag of chips:** Ibid.

77 **made too much money for the district:** Ibid.

78 **One in three kids in the United States over the age of four eats fast food every day:** "Fast Food Linked to Child Obesity," CBSNews.com, January 5, 2003, and S. A. Bowman, et al., "Effects of Fast-Food Consumption on Energy Intake and Diet Quality Among Children in a National Household Survey," *Pediatrics*, January 2004.

78 **enough calories to pack on six extra pounds each year:** Ibid.

78 **McDonald's hamburgers, Pizza Hut pizzas, or other brand-name fast foods:** "School Health Policies and Programs Study 2000," a report by the Centers for Disease Control and Prevention.

78 **ninety-five percent sold fast foods à la carte:** "2000 California High School Fast Food Survey," a report by Public Health Institute, February 2000.

78 **a Coca-Cola spokesperson described this arrangement as a "win-win-win":** Anna White, "Coke and Pepsi Are Going to School," *Multinational Monitor*, January 1999.

79 **warning children's doctors that obesity can be associated with:** "Soft Drinks in Schools," policy statement of the American Academy of Pediatrics, *Pediatrics*, January 2004.

79 **"The presence of a vending machine in a hall outside the classroom":** Ohio American Academy of Pediatrics Statement on Soft Drink Contracts in Schools.

79 **more than seventy-six percent of public schools in the United States sell soft drinks:** "School Health Policies and Programs Study 2000," a report by the Centers for Disease Control and Prevention.

79 **Kids can buy soda in nearly ninety-four percent of high schools:** "Healthy Schools for Healthy Kids," a report by the Robert Wood Johnson Foundation, 2003.

79 **In exchange for the exclusive right to sell their products:** Anne Ryman, "Schools Get Big Bucks in Soda Deals," *The Arizona Republic*, January 4, 2004.

79 **Parent Teacher Association accepted sponsorship by Coca-Cola:** Rhea R. Borja, "Coca-Cola Plays Both Sides of School Marketing Game," *Education Week*, November 5, 2003.

79 **a seat on the PTA's national board with full voting rights:** National PTA Board of Directors, John H. Downs Jr. bio.

79 **The PTA's defense is that Coca-Cola is endorsing them:** Rhea R. Borja, "Coca-Cola Plays Both Sides of School Marketing Game," *Education Week*, November 5, 2003.

80 **Atkins said it plans to help pay for the NEA's school health website:** Greg Toppo, "A Is For Atkins? Not Yet," *USA Today*, September 23, 2004.

80 **we bombard them with posters, banners, and scoreboard advertising:** Alex Molnar, "School Commercialism, Student Health, and the Pressure To Do More with Less," report from the Commercialism in Education Research Unit, Arizona State University, June 2003.

80 **"The question is, why are we harming these children?":** from a lecture by Alex Molnar in Berkeley, California, June 11, 2003.

81 **Soda partnerships are worth enormous amounts of money:** Anna White, "Coke and Pepsi Are Going to School," *Multinational Monitor*, January 1999.

81 **The groundwork for the school feeding frenzy was laid:** Democratic Staff of the Senate Committee on Agriculture, Nutrition, and Forestry, "Food Choices at School: Risks to Child Nutrition and Health Call for Action," May 18, 2004.

81 **Congress amended the 1966 Child Nutrition Act in 1970, allowing the USDA:** *National Soft Drink Association v. John R. Block*, Secretary Department of Agriculture, et al.

81 **"Profit had triumphed over nutrition":** Ibid.

81 **Following two years of hearings and 15,000 pages submitted in evidence:** Ibid.

81 **"can operate as a magnet for any child who inclines toward the non nutritious":** Ibid.

82 **"the Department of Agriculture was fully in the hands of the food industry":** From a phone interview with Carol Tucker Foreman, conducted by EH, June 2004.

82 **made $54.3 million in campaign contributions:** "Lobbyist Spending: Agribusiness" on the Center for Responsive Politics website.

82 **in 2000 the food and agriculture industries together spent $77.5 million lobbying:** "Agribusiness: Long-Term Contribution Trends," on the Center for Responsive Politics website.

82 **bills to redefine "foods of minimal nutritional value" sold in schools:** "Healthy Schools for Healthy Kids," a report by the Robert Wood Johnson Foundation, 2003.

83 **"Nobody stands up and says, 'We want kids to drink more soda'":** From a phone interview with Margo Wootan, conducted by EH, June 2004.

83 **At Samuel F. Gompers Continuation High School:** EH visited the high school in 2003 and 2004.

83 **buy the cheapest bulk foods, leaving out more expensive fresh produce:** "Healthy Schools for Healthy Kids," a report by the Robert Wood Johnson Foundation, 2003.

83 **"Schools are set up by our society as protected spaces":** From a speech by Alex Molnar to the American School Food Service Association, January 18, 2004.

83 **"It's a war zone in classrooms right now with regard to nutrition":** Ibid.

84 **which broadcasts overseas and to 12,000 schools in the United States:** From ChannelOne's website, ChannelOne.com.

84 **boasts an audience of eight million American sixth to twelfth graders:** Ibid.

84 **lends a television to each classroom in subscribing schools:** From a phone interview with Jim Metrock, conducted by EH, January 2004.

84 **for everything from M&M's to Snickers bars to McDonald's:** Ibid.

84 **Young "reporters" hawk products (such as newly released CDs) on the air:** From ChannelOne clips made available on Obligation's website.

84 **"government institutions, all luring our children to eat more Twinkies":** From a phone interview with Jim Metrock, conducted by EH, January 2004.

84 **He used to own a steel company:** Ibid.

84 **the states of California and New York have refused to let Channe One:** From Ibid. as well as the *Education Reporter*, "Channel One Still Plays to Captive Audiences," published by the Eagle Forum, December 1998. As evidence of how strange bedfellows can be, ultra-conservative Phyllis Schlafly, the founder of the Eagle Forum, has been an ardent critic of Channel One and other forms of school commercialism, joining many middle-of-the-road concerned parents as well as others who identify as liberals and progressives, including Ralph Nader.

84 **Others, such as the 20,000-student district in Shelby County, Alabama, have soured on the company and cancelled their contracts:** Ibid.

85 **"we ought to take them to the woodshed":** From a phone interview with Jim Metrock, conducted by EH, January 2004.

85 **the school board unanimously banned soft drinks:** Erika Hayasaki, "Schools to End Soda Sales," *Los Angeles Times*, August 28, 2002.

85 **"I have kindergartners who are ninety pounds at four or five years old":** Ibid.

85 **(although not in high schools) also as of 2004:** Nancy Vogel, "Davis to Sign Bills Limiting Soda Sales at School," *Los Angeles Times*, September 17, 2003.

85 **dismayed about students' easy access to soda at her school:** Jennifer Robertson and Cyndi Guerra Walter, "What It Took to Ban Soft Drinks in the LAUSD," *Lean Times*, newsletter of California Project Lean, November 2002.

85 **exclusive multimillion-dollar contract with Coke had to be kept secret:** Ibid.

86 **"very concerned about what the schools would do if they lost funding":** Ibid.

86 **It took a passionate three-hour debate:** Ibid.

86 **"money that schools are earning out of vending machines is very modest":** From phone interview with Margo Wootan, conducted by EH, June 2004.

86 **the state's schools *lose* $60 million in meal sales every year:** Democratic Staff of the Senate Committee on Agriculture, Nutrition, and Forestry, "Food Choices at School: Risks to Child Nutrition and Health Call for Action," May 18, 2004.

86 **Over the objections of the Texas Association of School Boards:** Peggy Fikac, "Texas Schools Put on Diet," *San Antonio Express-News*, March 4, 2004.

86 **Combs banned sales of candy, soda, and other sweets:** Texas School Nutrition Policy, Texas State Department of Agriculture.

87 **No soda sold in high school can be bigger than twelve ounces:** Ibid.

87 **only to sign an $8-million-a-year contract making Snapple the official drink:** David M. Herszenhorn, "New York Picks Its Beverage, for $166 Million," *New York Times*, September 10, 2003. Not only is Snapple the official drink of the city's schools, it's the official drink of New York City. Under the deal–which came to $126 million–Snapple has exclusive rights to the vending machine contracts not only in the city's 1,200 public schools but throughout city properties, including all city offices and police stations. In April 2004, New York City Comptroller William C. Thompson Jr. sued to block the pact with Snapple, accusing Mayor Michael Bloomberg of making a "backroom deal" and improperly awarding the contract to Snapple.

87 **Arkansas prohibits vending machines in elementary schools:** Darcia Harris Bowman, "States Target Vending Machines to Curb Child Obesity," *Education Week*, October 1, 2003.

87 **couldn't stop his district from signing a relatively small $150,000 contract:** From a phone interview with Rick Miriam, conducted by EH, February 2004.

87 **"Anything we do in the schools is putting a stamp of approval on it":** Ibid.

88 **"They should be held responsible. They shouldn't be allowed":** Ibid.

88 **Nixon cronies it was known as the western White House:** From a history of the Hyatt Newport Beach provided to guests.

88 **Campbell's, Tyson Foods, and Basic American Foods:** From the School Nutrition Association (formerly ASFSA) Fame 2004 awards brochure.

88 **a panel discussion on the school nutrition environment:** Town Hall Session: School Nutrition Environment, held at the School Nutrition Association Conference, January 19, 2004.

88 **Rather than suggest removing soda from the schools:** Ibid.

89 **there followed an uncomfortable silence:** Ibid.

89 **and make up only approximately one percent of the group's 53,303 members:** From email interview with Donna Wittrock, conducted by EH, May 2004.

89 **the food industry contributes about $425,000 in dues and sponsorships:** Ibid.

89 **"The organization is not beholden to the food and beverage industry":** Ibid.

89 **the black-haired, boyish-looking Cliff Medney:** Point-Counterpoint: Marketing to Children–Two Opinions, a panel with Cliff Medney and Alex Molnar at the School Nutrition Association conference, January 18, 2004.

89 **they were in a unique position to market good nutritional messages:** Ibid.

90 **"If they're informed and intelligent and smart, then kids":** Ibid.

90 **but he warned them not to be wooed by marketers:** Ibid.

90 **He mused over whether, like an offer made to one of the Marx brothers:** Ibid.

90 **"I like partnerships. Nobody doesn't like a good partner":** Ibid.

90 **contends that soda is part of a balanced diet that aids hydration:** "NSDA Statement on Efforts to Ban or Restrict the Sale of Carbonated Soft Drinks in Schools," National Soft Drink Association, from the website nsda.org.

90 **put their soda profits into physical education:** "The Value We Bring," National Soft Drink Association, from the website nsda.org.

91 **"How far do you have to run to run off two twenty-ounce Big Gulps":** Point-Counterpoint: Marketing to Children–Two Opinions, a panel with Cliff Medney and Alex Molnar at the School Nutrition Association conference, January 18, 2004.

91 **drastic cuts in elementary and high school physical education:** "School Health Policies and Programs Study 2000," Centers for Disease Control and Prevention, 2000.

91 **children get less than thirty minutes of exercise:** M. V. Chakravarthy and F. W. Booth, "Inactivity and Inaction: We Can't Afford Either," *Archives of Pediatrics and Adolescent Medicine*, August 2003.

91 **high school students taking daily physical education classes declined more than thirty percent:** Neil Sherman, "Parents Want Schools to Fight Fat," *HealthScout*, September 13, 2000.

91 **Even though the CDC recommends requiring daily PE:** "Promoting Better Health for Young People Through Physical Activity and Sports: A Report to the President," CDC, Fall 2000.

92 **"First of all, let me say, 'People, get your facts straight!'":** Tom Bachmann, "Why Are Our Kids Getting Fat?" *Beverage Industry*, May 2002.

92 **"Enough is enough: What is the plan?":** Scott Leith, "Coca-Cola Enterprises Defends Presence in Schools," *Atlanta Journal and Constitution*, April 6, 2003.

92 **he asked his public affairs director:** Ibid. At least part of the plan became clear when two months after Alms posed the question to Downs, Downs himself joined the national board of the Parent Teacher Association, at the same time that Coke became a sponsor of the PTA. Neither the PTA nor Coca-Cola has disclosed what the sponsor-

ship cost. When news of Downs's June appointment reached consumer groups in September 2003, they quickly condemned it. Gary Ruskin, executive director of Commercial Alert, called the PTA's intimate association with Coca-Cola a "massive conflict of interest," the *New York Times* reported on September 3, 2003. Others branded it a desperate effort on Coke's part to keep its soda in the schools.

92 **Coca-Cola Enterprises quickly decided on a strategy:** Ibid.

92 **also made it a policy to offer a choice of bottled water and juices:** Ibid.

92 **"The school system is where you build brand loyalty":** Ibid.

92 **Pennsylvania State University signed a $14 million:** Anna White, "Coke and Pepsi Are Going to School," *Multinational Monitor*, January 1999.

92–93 **the most encompassing school deal with the beverage industry:** Ibid.

93 **Other colleges quickly followed:** Ibid.

93 **it wasn't until 1998 that exclusive deals between school districts and Big Soda:** Ibid.

93 **that he called himself the "Coke Dude":** Ibid.

94 **"yet we infuse [students] with megadoses of sugar. We probably shouldn't be doing that":** "Virginia Schools' Dilemma: Profit or Diet?" Associated Press, April 12, 2004.

94 **a whopping ninety-two percent of teachers and ninety-one percent of parents:** "Healthy Schools for Healthy Kids," a report by the Robert Wood Johnson Foundation, 2003.

95 **there hasn't been a vending machine on campuses in sixteen years:** Ibid.

95 **a former restaurant manager, substituted fast food with pasta, rice bowls:** Chris Woolston, "School Lunches: Invasion of the Body Fatteners," *Consumer Health Interactive*, July 24, 2002.

95 **kids were drinking something called "Koolies":** From in-person interview with Beverly Girard, conducted by EH, January 2004.

95 **"People thought they were doing the right thing before I got there":** Ibid.

96 **She added more fresh fruits and vegetables, although they take more work:** Ibid.

96 **"but in many cases it's not coming into the family home":** Ibid.

96 **Each time the food improved, the cafeterias made more money:** Ibid.

96 **"They're not lying that they need the money. It's just the wrong way":** Ibid.

96 **"I almost laid my body down at a school board meeting. I almost quit":** Ibid. Since that interview, Girard said in an email that some school board members were having second thoughts about the district's contract with Coca-Cola. The contract is up for renewal in May 2005.

96 **Al Schieder was as disturbed by the cafeteria food:** From a phone interview with Al Schieder, conducted by EH, May 2004.

96 **"I realized we had two high schools. One was affluent. One was poor":** Ibid.

96 **the food service program was losing $200,000 a year:** Ibid.

97 **He put garden bars in every elementary school:** Ibid.

97 **Every lunch is freshly made and geared toward the tastes:** Ibid.

97 **But his sales more than doubled from $1.7 million to more than $3.5 million:** From a talk by Al Schieder given at Condition Critical: Covering Children's Health, a conference by the Casey Journalism Center on Children and Families, March 2004.

97 **"We can wallpaper schools with (posters) about what nutrition":** Ibid.

98 **Aptos is the city's most racially diverse school:** From an in-person interview with Linal Ishibashi, conducted by EH, January 2004.

98 **forty percent of the students come from families with incomes low enough:** Ibid.

98 **The school cafeteria is a plain brown and beige room:** From a visit to Aptos Middle School by EH, January 2004.

98 **Ishibashi asked parent and active school volunteer Dana Woldow:** From an in-person interview with Linal Ishibashi, conducted by EH, January 2004.

98 **she had simply dismissed the fare at Aptos as "garbage":** From in-person interview with Dana Woldow, conducted by EH, January 2004.

98–99 **"There are many ways of combating it, and you need all of them":** Ibid.

99 **Woldow bypassed the nutrition department that had refused to give up chips:** Ibid.

99 **The parents found the product contained not only several forms of sugar:** "Junk Food Out, Profits in at San Francisco Middle School," EducationNews.org.

99 **mega-colossal cheeseburgers (fifty-eight percent fat), hot links (seventy-seven percent fat):** Ibid.

99 **They found a vendor who would make them sushi with no MSG:** From in-person interview with Dana Woldow, conducted by EH, January 2004.

99 **The California roll sushi (no raw fish) wouldn't be out of place:** From a visit to Aptos Middle School by EH, January 2004.

100 **"I think it's good. It's really tasty":** Ibid.

100 **"You ever get your kids to try something new?":** From in-person interview with Dana Woldow, conducted by EH, January 2004.

100 **turned a profit for the first time, and it has remained profitable:** Data from the San Francisco Unified School District Student Nutrition office as reported in "Waistlines and Bottom Lines," a report from the Aptos Student Nutrition Committee.

100 **School District began a district-wide "no empty calories" policy:** "Junk Food Off the Menu at San Francisco Schools," news release from Parents Advocating School Accountability, February 4, 2004.

100 **"The most crowded time at the counselor's office used to be right after lunch":** From in-person interview with Dana Woldow, conducted by EH, January 2004.

101 **students at Aptos scored forty-five points higher:** "2002–2003 Academic Performance Index (API) Growth Report," California Department of Education, revised June 8, 2004.

101 **academic achievement is strengthened when students are well nourished:** Studies cited in *Taking Action for Healthy Kids: A Report by Action for Healthy Kids* on the Healthy Schools Summit held October 2002 (Center on Hunger, Poverty and Nutrition Policy, 1995; President's Council, 1999).

101 **"Now I find kids running up to me regularly, saying":** From in-person interview with Linal Ishibashi, conducted by EH, January 2004.

101 **"If I went to the district, and I said that for $50,000 I can sell you a program":** From in-person interview with Dana Woldow, conducted by EH, January 2004.

Chapter 5: The Sedentary Bunch

Page

103 **Another 86,000 square feet of gambling space:** Vegas.com.

103 **1,500 hotel rooms:** Adam Steinhauer, "Grand, Stratosphere Divorce Final," *Las Vegas Review-Journal*, October 10, 1997.

103 **complex cost $550 million:** Ibid.

104 **looks like a typical Las Vegas elementary school with its crowded blacktop:** CW visited Park-Edison elementary in May 2004.

104 **handed over John S. Park and five other elementary schools to Edison Schools, Inc.:** Lisa Kim Bach, "Edison Contract: Seven Schools Sublet," *Las Vegas Review Journal*, April 7, 2001.

104 **a company that specializes in "rescuing" failing schools:** Brian O'Reilly, "Why Edison Doesn't Work," *Fortune*, December 9, 2002.

104 **some even staged mock funerals:** "Editorial: The Edison 'Funeral,'" *Las Vegas Review-Journal*, August 26, 2001.

104 "**wear the same shirt day after day":** From in-person interview with Roy Leon, conducted by CW, May 2004.

105 **emphasizes physical activity above all else:** Kathy Slobogin, "New Phys Ed Favors Fitness Over Sports," CNN.com, May 17, 2001.

106 **"four major sports aren't the only form of exercise":** From in-person interview with Roy Leon, conducted by CW, May 2004.

106 **spending per student falls about $2,000 short of the national average:** "NEA Study Finds Investment in U.S. Public Schools Lagging as Education Needs Rise," National Education Association press release, May 21, 2003.

106 **the state doesn't require a single minute of physical education for grades kindergarten through eighth:** From phone interview with Cathy Altenburg (president of Eastern District of Nevada Alliance for Health, Physical Education, Recreation, and Dance), conducted by CW, June 2004.

107 **only one fourth to one third require it for middle school kids:** "School Health Policies and Programs Study 2000," Centers for Disease Control and Prevention, 2000.

107 **leave the task up to regular classroom teachers:** Ibid.

107 **the average gym class has more than forty students:** "Cuts Leave More and More Public School Children Behind," National Education Association, December 2003 January 2004.

107 **a single class can have as many as seventy kids:** Refers to Fay Herron Elementary.

107 **the number of kids *in one class* can reach triple digits:** Refers to Bellaire Elementary.

107 **"PE teachers are very worried":** From phone interview with Charlene Burgeson, conducted by CW, May 2004.

107 **laid off all five PE teachers in the elementary schools:** From phone interview with Stephanie Perkins, conducted by LU, June 2004.

107 **"I'm not going to be able to do it":** From phone interview with Cheryl Madsen, conducted by LU, June 2004.

108 **"I think the priorities are a mess":** From phone interview with Stephanie Perkins, conducted by LU, June 2004.

108 "**truly keep his insulin down and glucose under control with activity":** From phone interview with Becky Kopecky, conducted by LU, June 2004.

108 **"Maybe we're even contributing to the epidemic":** From phone interview with Kathleen Sanders, conducted by LU, June 2004.

109 **"They grew up to be principals":** From a lecture and demonstration of the New PE by Susan Kogut at the Casey Journalism Center, March 2004.

109 **with the passage of the No Child Left Behind Act of 2001:** From a lecture by William Dietz at the 2004 annual conference of the American Alliance for Health, Physical Education, Recreation, and Dance, New Orleans, March 2004.

109 **penalizes schools whose students do not score well:** "Cuts Leave More and More Public School Children Behind," National Education Association, December 2003 January 2004.

109 **"We're collecting stories from around the country":** From phone interview with Daniel Kaufman, conducted by CW, June 2004.

110 **don't offer regular recess:** "School Health Policies and Programs Study 2000," Centers for Disease Control and Prevention, 2000.

110 **"A lot of kids get a few minutes after lunch, and that's it":** From phone interview with Melinda Sothern, conducted by CW, December 2001.

110 **"a very shortsighted approach":** From phone interview with James Sallis, conducted by CW, June 2004.

110 **were actually *less* active after school on days when they don't have PE:** D. Dale and C. Corbin, "Restricting Opportunities to Be Active During School Time: Do Children Compensate by Increasing Physical Activity Levels After School?" *Research Quarterly for Exercise and Sport,* September 2000.

110 **under siege at the worst possible time:** From lecture by David Satcher at the 2004 annual conference of the American Alliance for Health, Physical Education, Recreation, and Dance, New Orleans, March 2004.

111 **It's no mere coincidence:** A. Cotton, "Report Card: State Efforts to Control Obesity," news release, University of Baltimore.

111 **"People would say this is kind of a no-brainer– of course kids should get recess":** From phone interview with Sally Harrell, conducted by LU, June 2004.

111 **"I thought that's what home is for":** From phone interview with Ruel Parker, conducted by LU, June 2004.

112 **"between two and five minutes between classes":** From phone interview with Deanna Ryan, conducted by LU, June 2004.

112 **twenty-five percent of kindergartners in Georgia–who are mandated to have a full day of school–do not have daily recess:** From phone interview with Olga Jarrett, conducted by LU, June 2004.

112 **included parents, teachers, and the local chapter of the American Academy of Pediatrics:** From phone interview with Deanna Ryan, conducted by LU, June 2004.

112 **"cooped-up kids to get their bodies under control long enough to focus on anything"**: Michelle Gregg, written testimony in favor of bill.

112–113 **she collected 400 signatures in protest:** From phone interview with Anne Torrez, conducted by LU, June 2004.

113 **"if our kids can't have recess":** Ibid.

113 **"And all the other kids seem less wild when we do have recess":** From phone interview with Cydney Torrez, conducted by LU, June 2004.

113 **"which might help them become healthier adults":** From phone interview with Mark Taylor, conducted by LU, June 2004.

113 **"should not be reduced or replaced with academic classes":** "Childhood Obesity," American Obesity Association, 2000.

113–114 **"close the gap between our science and our policies":** From lecture by David Satcher at the 2004 annual conference of the American Alliance for Health, Physical Education, Recreation, and Dance, New Orleans, March 2004.

114 **the most sedentary in history:** Julie Yoshioka, "Thinking Outside the Box," *UCLA Daily Bruin*, April 24, 2000.

114 **get less than thirty minutes of exercise each day:** M. V. Chakravarthy and F. W. Booth, "Inactivity and Inaction: We Can't Afford Either," *Archives of Pediatrics and Adolescent Medicine*, August 2003.

114 **at least one hour of moderate to vigorous activity every day:** "Children Need Greater Amount of Physical Activity in 2004," National Association of Sports and Physical Education, December 31, 2003.

114 **hearts would be stronger:** Theodore Ganley, "Exercise and Children's Health," *The Physician and Sports Medicine*, February 2000.

114 **much less likely to put on weight:** Ibid.

114 **more capable, confident, and energetic:** From phone interview with James Sallis, conducted by CW, June 2004.

114 **relieves the symptoms of anxiety and depression:** From phone interview with James Sallis, conducted by CW, June 2004. Also Theodore Ganley, "Exercise and Children's Health," *The Physician and Sports Medicine*, February 2000.

114 **"the biggest motivation for kids to exercise":** From phone interview with James Sallis, conducted by CW, June 2004.

114 **allot more time to physical education and recess:** Ibid.

114 **"you make a commitment, and you have the work ethic to see it through":** From lecture by Alice Mehaffey at the 2004 annual conference of the American Alliance for Health, Physical Education, Recreation, and Dance, New Orleans, March 2004.

115 **flanked by two cornfields and a pig farm:** From phone interview with Tammy Brant, conducted by CW, June 2004.

115 **"can't even walk to school because they'd get run over by a tractor":** From lecture by Tammy Brant at the 2004 annual conference of the American Alliance for Health, Physical Education, Recreation, and Dance, New Orleans, March 2004.

115 **almost all parents have full-time jobs:** From phone interview with Tammy Brant, conducted by CW, June 2004.

115 **"kids don't have the opportunity to do things":** Ibid.

115 **$500 million marketing campaign to launch its Xbox gaming system:** "Ed Fries Discusses Xbox Launch," gamespot.com, November 14, 2001.

116 **"young kids are just naturally active":** From phone interview with Judy Young, conducted by CW, November 2001.

116 **raises the lifetime risk of at least seventeen chronic conditions:** F. W. Booth, et al., "Waging War on Modern Chronic Diseases: Primary Prevention Through Exercise Biology," *Journal of Applied Physiology*, 2000.

116 **including type 2 diabetes, heart disease, arthritis, several types of cancer, and, of course obesity:** Ibid. See also "Physical Activity Fundamental to Preventing Disease," U.S. Department of Health and Human Services, June 20, 2002.

116 **may develop complications such as high blood pressure, high cholesterol, or insulin resistance:** M. V. Chakravarthy and F. W. Booth, "Inactivity and Inaction: We Can't Afford Either," *Archives of Pediatrics and Adolescent Medicine*, August 2003. The authors use the term "sedentary death syndrome" to refer to the deadly complications of inactivity. The authors also note that "sedentary death syndrome has its genesis during childhood and adolescence."

116 **can be enough to put a target on a child's back:** From a lecture by Timothy Lohman at the 2004 annual conference of the American Alliance for Health, Physical Education, Recreation, and Dance, New Orleans, March 2004.

116 **"we'd lower the risk of many diseases":** Ibid.

116 **believe that inactivity is even more dangerous than obesity:** From a lecture by Stephen Blair at the 2004 annual conference of the American Alliance for Health, Physical Education, Recreation, and Dance, New Orleans, March 2004.

117 **"Physical activity adds little to weight reduction":** From a lecture by William Dietz at the 2004 annual conference of the American Alliance for Health, Physical Education, Recreation, and Dance, New Orleans, March 2004.

117 **a crucial component of any successful long-term weight loss program:** Ibid.

117 **by simply walking an extra ten minutes each day:** M. V. Chakravarthy and F. W. Booth, "Inactivity and Inaction: We Can't Afford Either," *Archives of Pediatrics and Adolescent Medicine*, August 2003.

117 **and spend at least thirty of those minutes walking:** Ibid.

117 **excellent chance to read or catch up on homework:** Ibid. Reading and homework are not culprits in the epidemic of inactivity. Researchers at the University of Montreal found that kids who spend a lot of time on these "productive sedentary behaviors" are actually more active than other kids. (D. E. Feldman, et al., "Is Physical Activity Differentially Associated with Different Types of Sedentary Pursuits?" *Archives of Pediatrics and Adolescent Medicine*, August 2003.)

117 **"but things haven't improved":** From a lecture by Robert Pangrazi at the 2004 annual conference of the American Alliance for Health, Physical Education, Recreation, and Dance, New Orleans, March 2004.

118 **it all counts, and it all adds up:** See "A New View of Physical Activity," Centers for Disease Control and Prevention, November 17, 1999. As strange as it may seem, even playing video games may be better than nothing. In 1991, William Dietz and a colleague published a study showing that video games increased a child's resting metabolic rate by eighty percent.

119 **recently tracked the physical activity levels of more than 2,000 girls:** S. Kim, et al., "Decline in Physical Activity in Black Girls and White Girls During Adolescence," *The New England Journal of Medicine*, September 5, 2002. A few other trends stood out. At all ages, the thinner girls tended to get more exercise than the heavier girls. Girls whose parents were well educated were also especially likely to stay active, possibly because their parents understood the many benefits of an active lifestyle. On the other hand, household income didn't really seem to matter; girls from poor families were just as active (or inactive) as girls from rich families.

119 **three to four-and-a-half hours of television every day:** M. V. Chakravarthy and F. W. Booth, "Inactivity and Inaction: We Can't Afford Either," *Archives of Pediatrics and Adolescent Medicine*, August 2003.

119 **were more than four times more likely than other kids to be overweight:** S. L. Gortmaker, et al., "Television Viewing as a Cause of Increasing Obesity Among Children in the United States, 1986–1990," *Archives of Pediatrics and Adolescent Medicine*, April 1996.

119 **"biggest single risk factor for becoming overweight":** From phone interview with James Sallis, conducted by CW, June 2004.

119–120 **get just as much overall physical activity as those who watch very little:** D. E. Feldman, et al., "Is Physical Activity Differentially Associated with Different Types of Sedentary Pursuits?" *Archives of Pediatrics and Adolescent Medicine*, August 2003.

120 **intertwined with the junk food habit:** From phone interview with James Sallis, conducted by CW, June 2004. See also "Inactivity and Inaction: We Can't Afford Either," *Archives of Pediatrics and Adolescent Medicine*, August 2003.

120 **gaining excess weight over the next year:** J. O'Loughlin, et al., "One- and Two-year Predictors of Excess Weight Gain Among Elementary Schoolchildren in Multiethnic, Low-Income, Inner-City Neighborhoods," *American Journal of Epidemiology*, October 15, 2000.

120 **but there's some truth in those tall tales:** From phone interview with Marya Morris, conducted by CW, January 2004.

121 **between school and home only about thirteen percent of the time:** "Walk to School Programs Fact Sheet," Centers for Disease Control and Prevention, December 2002.

121 **was taking kids to school:** Bruce Appleyard, "Safe Routes to School: Introduction," National Center for Bicycling and Walking.

121 **Violent crimes against children aren't really more common today than they were twenty years ago:** E. J. Mundell, "Report Gives Mixed Grades on Kids' Health," HealthDayNews, March 24, 2004.

121 **typically built on the biggest, cheapest lots available:** From phone interview with Marya Morris, conducted by CW, January 2004.

121 **sprawl puts up other barriers to exercise:** Ibid.

122 **"onto the thoroughfare to get to the school":** From in-person interview with Joan Miller, conducted by EH, May 2004.

122 **adds an average of six pounds to every adult:** R. Ewing, et al., "Relationship Between Urban Sprawl and Physical Activity, Obesity, and Morbidity," *American Journal of Health Promotion*, September/October 2003.

122 **"left in the dust":** From in-person interview with Molly and Libby Markus, conducted by CW, May 2004.

122 **have deep sympathy for heavy kids:** From a lecture by Robert Pangrazi at the 2004 annual conference of the American Alliance for Health, Physical Education, Recreation, and Dance, New Orleans, March 2004.

124 **"It's a painful experience, and they learn to hate it":** From a lecture by Robert Pangrazi at the 2004 annual conference of the American Alliance for Health, Physical Education, Recreation, and Dance, New Orleans, March 2004.

124 **get moving with the right encouragement:** From phone interview with James Sallis, conducted by CW, June 2004.

124 **a throng of kids gathers in the central courtyard:** CW visited Fay Herron in May 2004.

125 **the largest elementary school in Nevada:** Information provided by Fay Herron Elementary.

125 **"I have to change their attitude":** From in-person interview with Jurgen Kraehmer, conducted by CW, May 2004.

126 **"A lot of kids have given up on themselves":** Ibid.

127 **designed to encourage nine- to thirteen-year-olds to become more physically active:** CDC's VERB website, http://www.cdc.gov/youthcampaign/.

128 **the first such campaign ever to get adequate funding:** From a lecture by William Dietz at the 2004 annual conference of the American Alliance for Health, Physical Education, Recreation, and Dance, New Orleans, March 2004.

128 **kids stream outside:** Ads can be viewed at the VERB website, http://www.cdc.gov/youthcampaign/.

128 **enough to prevent a weight gain of five pounds:** James Sallis, "Talk Today: Youth Exercise," USAToday.com, August 22, 2002.

128 **put up stoplights, and build pedestrian bridges:** Bruce Appleyard, "Safe Routes to School: Case Studies," National Center for Bicycling and Walking.

128 **where parents walk to school with groups of neighborhood kids:** "The Walking School Bus: Bringing Schools and Communities Together to Create Safer Walking Environments for Kids," National Highway Traffic Safety Administration.

128 **walked or rode their bikes to school increased by eighty percent:** Bruce Appleyard, "Safe Routes to School: Case Studies," National Center for Bicycling and Walking.

128 **James Oberstar (D–MN) in the U.S. House and Jim Jeffords (I–VT) in the Senate:** From phone interview with Martha Roskowski, conducted by CW, May 2004.

128 **version of the bill would allot $420 million over six years:** From phone interview with Martha Roskowski, conducted by CW, May 2004.

129 **won't come close to giving every kid a decent route to school:** Ibid.

129 **"federal grant money to get people excited about a project":** Ibid.

129 **urban planners have been looking for better ways to build communities:** From phone interview with Marya Morris, conducted by CW, January 2004.

129 **"new level of interest in the design of our communities":** From phone interview with Martha Roskowski, conducted by CW, May 2004.

130 **"people with MDs backing us up makes a huge difference":** From phone interview with Marya Morris, conducted by CW, January 2004.

130 **Government agencies such as the CDC:** A description of the CDC's Active Community Environments Initiative can be found at http://www.cdc.gov/nccdphp/dnpa/aces.htm.

130 **$70 million to promoting smart growth and active communities:** R. Ewing et al., "Relationship Between Urban Sprawl and Physical Activity, Obesity, and Morbidity," *American Journal of Health Promotion*, September/October 2003.

130 **expect to attract 30,000 residents in the next decade:** M. Larkin, "Can Cities Be Designed to Fight Obesity?" *The Lancet*, September 27, 2003.

130 **state that already boasts the nation's thinnest residents:** "Colorado Still Lean State, But Obesity Increased Over Decade," Colorado Department of Public Health and Environment, June, 12, 2003.

131 **real change started at home:** From in-person interview with Molly and Libby Markus, conducted by CW, May 2004.

Chapter 6: Obesity Goes Global

Page

133 **autumn afternoon on New Zealand's North Island:** CW visited New Zealand in March 2003. This scene took place at the Goat Island Marine Reserve north of Auckland.

134 **and ten percent were extremely heavy:** "NZ Food, NZ Children: Findings of the 2002 National Children's Nutrition Survey," November 2003.

134 **nobody knows how much things have changed over the years:** From phone interview with John Birkbeck, conducted by CW, January 2004.

134 **Egyptian preschoolers who are overweight nearly quadrupled:** Cara Ebbeling, et al., "Childhood Obesity: Public Health Crisis, Common Sense Cure," *The Lancet*, August 2002.

134 **nearly tripled between 1984 and 1994:** Ibid.

134 **overweight Scottish boys and girls roughly doubled:** Ibid.

134 **Australian girls ages seven to fifteen grew nearly fivefold:** Ibid.

134 **Preschoolers in Ghana, teenagers in Brazil:** Ibid.

135 **a country where only one in one thousand preschoolers is overweight:** Mercedes de Onis and Monika Blossner, "Prevalence and Trends of Overweight Among Preschool Children in Developing Countries," *American Journal of Clinical Nutrition*, 2000.

136 **"personal responsibility" is the key to battling the obesity epidemic worldwide:** Rob Stein, "U.S. Says It Will Contest WHO Plan to Fight Obesity But Claim of Faulty Science Is Rejected by Nutritionists," *Washington Post*, January 16, 2004.

136 **"That's a very American approach":** From a phone interview with Neville Rigby, conducted by CW, February 2004.

136 **"it's not our food, it's personal responsibility":** From a phone interview with Michael Lowe, conducted by CW, February 2004.

137 **ten-year-olds in the United Kingdom and thirty percent of ten-year-olds in Munich, Germany, are overweight:** "About Obesity," Donald B. Brown Research Chair on Obesity, Laval University, Quebec, Canada.

137 **twenty percent were overweight by CDC standards:** M.-F. Rolland-Cachera, et al., "Body Mass Index in 7–9-Y-Old French Children: Frequency of Obesity, Overweight, and Thinness," *International Journal of Obesity*, 2002.

137 **on a par with the United States in the late 1980s, a precarious position to be sure:** Ibid.

137 **in northern countries such as Sweden (eighteen percent), Denmark (fifteen percent), and Germany (sixteen percent):** T. Lobstein, et al., "Obesity in Children and Young People: A Crisis in Public Health," *Obesity Reviews*, vol. 5, 2004.

137 **tend to be heavier than people in the north, says Italian obesity expert Francisco Branca, PhD:** From phone interview with Francisco Branca, conducted by CW, January 2004. See also T. Lobstein, et al., "Obesity in Children and Young People: A Crisis in Public Health," *Obesity Reviews*, Vol. 5, 2004.

137 **why obesity rates change with latitude:** Ibid.

137 **but Branca sees a corresponding gradient in feelings about exercise:** From phone interview with Francisco Branca, conducted by CW, January 2004.

137 **a six-year span of severe economic hardship throughout the country:** T. Lobstein, et al., "Obesity in Children and Young People: A Crisis in Public Health," *Obesity Reviews*, vol. 5, 2004.

138 **another impoverished country–Lithuania–had the fewest overwieght kids:** Inge Lissau, et al., "Body Mass Index and Overweight in Adolescents in 13 European Countries, Israel, and the United States," *Archives of Pediatrics and Adolescent Medicine*, January 2004.

138 **preschoolers are overweight and nearly that many are wasting away from malnutrition:** Mercedes de Onis and Monika Blossner, "Prevalence and Trends of Overweight Among Preschool Children in Developing Countries," *American Journal of Clinical Nutrition*, vol. 72, 2000.

138 **seven percent of preschool kids are overweight in Malawi:** Ibid.

138 **annual per capita income less than $200 (US):** "Malawi Profile: Economy," Nationmaster.com.

138 **a village in rural China, for instance, and you can go a long time without seeing a kid with a chubby face:** CW visited China in July 2004. See also William Scott Pappert, "Obesity in China: Tracking Nutrition in Transition," University of Pittsburgh.

138 **incredible upsurge in childhood obesity in Beijing and other Chinese cities during the last decade-and-a-half:** J. Luo and F. B. Hu, "Time Trends of Obesity in Pre-School Children in China from 1989 to 1997," *International Journal of Obesity*, vol. 26, 2002.

138 **increased more than sixfold to nearly thirteen percent:** Ibid.

139 **childhood obesity in Chinese cities typifies a global trend:** From phone interview with Monika Blossner, conducted by CW, February 2004.

139 **sweetened with sugars, from cereals to white bread to potato chips:** From phone interview with Barry Popkin, conducted by CW, May 2004.

139 **average person on the planet got seventy-four more calories each day from added sweeteners in 2000 than in 1962:** Barry Popkin and Samara Nielsen, "The Sweetening of the World's Diet," *Obesity Research*, November 2003.

139 **attributes some of that increase to good marketing:** From phone interview with Barry Popkin, conducted by CW, May 2004.

139 **boost in calories from sweeteners was tightly tied to growth of cities:** Barry Popkin and Samara Nielsen, "The Sweetening of the World's Diet," *Obesity Research*, November 2003.

140 **the sweetening of the world's diet is a significant factor in the fattening of the world's children:** From email interview with Barry Popkin, conducted by CW, May 2004.

140 **affectionate term for overweight children–*xiao pangzi*, or "little fatties":** Seth Mydans, "Clustering in Cites, Asians Are Becoming Obese," *New York Times*, March 13, 2003.

140 **sign of family wealth and good fortune:** Ibid.

140 **adopted its one-child law in 1979, the doting has grown even more intense:** William Scott Pappert, "Obesity in China: Tracking Nutrition in Transition," University of Pittsburgh.

140 **one hundred million people have high blood pressure and twenty-six million have diabetes:** Duncan Hewitt, "China Battles Obesity," BBC News, May 23, 2000.

140 **"fat camps" that have sprung up around Chinese cities:** Ibid.

140–141 **lamented the state of Chinese children in a broadcast on National Public Radio:** Reprinted in William Scott Pappert, "Obesity in China: Tracking Nutrition in Transition," University of Pittsburgh.

141 **eating fewer vegetables and less rice but more red meat, eggs, dairy products, sodas, and sugars:** Ibid. See also Duncan Hewitt, "China Battles Obesity," BBC News, May 2000.

141 **boosting their intake of fats and calories and almost certainly expanding their waistlines:** William Scott Pappert, "Obesity in China: Tracking Nutrition in Transition," University of Pittsburgh. See also Duncan Hewitt, "China Battles Obesity," BBC News, May 2000.

141 **supporting the local sugar and soda industries with loans and hefty subsidies:** William Scott Pappert, "Obesity in China: Tracking Nutrition in Transition," University of Pittsburgh.

141 **greatly increased the availability and lowered the price of sodas and sweets for Chinese families:** Ibid.

141 **Starbucks, KFC, Pizza Hut, and (it probably goes without saying) McDonald's:** James L. Watson, "China's Big Mac Attack," *Foreign Affairs*, May/June 2000.

141 **It's not even seen as an American restaurant:** Ibid.

141 **two things often lacking in traditional restaurants:** Ibid.

141 **"leisure centers for seniors and after-school clubs for students":** Ibid.

142 **closing restaurants in other parts of the world:** "China Has a Big Mac Attack," BBC News, November 13, 2002.

142 **one hundred new restaurants spring up in China every year:** Ibid.

142 **company aims to have 1,000 Chinese outlets in time for the summer Olympics in 2008:** "McDonald's Bets on Chinese Growth," BBC News, February 26, 2004.

142 **including McDuck's, Mordornal, and Mcdonald's:** James L. Watson, "China's Big Mac Attack," *Foreign Affairs*, May/June 2000.

142 **happen to be the cheapest on the planet:** "China Has the Cheapest Big Macs," Reuters, April 30, 2003.

142 **it's called the "Big Mac Index":** Ibid.

142 **typically cost about $1.20 in U.S. dollars, less than half the price in the United States:** Ibid.

142 **"undervaluation" of the Chinese yuan:** Ibid.

143 **vastly outnumbered by sidewalk vendors tending steaming pots of traditional fare: rice mixed with chickpeas, carrots, and small bits of lamb:** From phone interview with Francisco Branca, conducted by CW, January 2004.

143 **kids on scales, investigate their diets, and unravel this mystery:** Ibid.

143 **other children are starting to get the first regular meals of their lives:** Ibid.

143 **"families eat meat with practically every meal, including the fat":** Ibid.

143 **they represent the future of many developing countries:** "Fighting Hunger Today Could Fight Obesity Tomorrow," UN Food and Agriculture Organization, February 11, 2004.

143–144 **millions of children and infants suffer from malnutrition:** "Moderate Malnutrition Kills Millions of Children Needlessly," Cornell University News Service, July 1, 2003.

144 **missing out on vital micronutrients such as iron, zinc, and iodine:** From phone interview with Francisco Branca, conducted by CW, January 2004.

144 **their metabolism slows down:** Ibid.

144 **progressively weaker and become more and more vulnerable to infectious diseases:** "Moderate Malnutrition Kills Millions of Children Needlessly," Cornell University News Service, July 1, 2003.

144 **"reprograms" their bodies to put on the most amount of fat with the least amount of calories:** "Fighting Hunger Today Could Fight Obesity Tomorrow," UN Food and Agriculture Organization, February 11, 2004. Also from phone interview with Francisco Branca, conducted by CW, January 2004.

144 **often develop the same type of "metabolic syndrome":** From phone interview with Francisco Branca, conducted by CW, January 2004. "Metabolic Syndrome" refers to heart disease risk factors that often occur together, including extra weight around the midsection, high cholesterol, high blood pressure, and insulin resistance. ("Metabolic Syndrome," American Heart Association, 2004.)

144 **that afflicts American teenagers with a love for fast food:** "One Million American Teenagers Already Facing Heart Disease," International Obesity Task Force, January 24, 2004.

144 **soon send the global epidemic of obesity into high gear:** "Fighting Hunger Today Could Fight Obesity Tomorrow," UN Food and Agriculture Organization, February 11, 2004.

144 **could cripple the healthcare budgets of many poor countries:** Ibid.

144 **prevent malnutrition in pregnant women and young children:** Ibid.

145 **example of a country with every possible advantage in the fight against obesity:** Deborah Ball, "Swedish Kids Show Difficulty of Fighting Fat," *Wall Street Journal*, December 2, 2003.

145 **nearly as much TV as their American peers:** From phone interview with Claude Marcus, conducted by CW, January 2004.

145 (**Swedish television prohibits ads aimed at children.**): Ibid. Also Deborah Ball, "Swedish Kids Show Difficulty of Fighting Fat," *Wall Street Journal*, December 2, 2003.

145 **walk or bike to school:** From phone interview with Claude Marcus, conducted by CW, January 2004.

145 **healthy lunches when they get there:** Ibid.

145 **regular, vigorous PE classes:** Ibid.

145 **overweight seven-year-olds in Stockholm increased from eight percent in 1989 to approximately twenty percent in 2004:** Ibid.

145 **many 200-pound kids to believe in this romanticized ideal:** Ibid.

145 **are the ones most likely to show up in his office:** Ibid.

145 **radically changed their approach to eating:** Ibid.

146 **new types of leisure-time fun:** Ibid.

146 **kids today don't seem much different from kids twenty years ago:** CW lived in New Zealand in 1985 and visited again in 2003.

146 **"Kids are strongly influenced by American culture":** From phone interview with Lorna Gillespie conducted by CW, January 2004.

146–147 **every New Zealand town had at least one "takeaway" shop:** Observed by CW, 1985.

147 **have recently stolen many customers from the takeaway shops:** From phone interview with John Birkbeck, conducted by CW, January 2004.

147 **"now there are a whole multitude of other temptations":** From phone interview with Celia Murphy, conducted by CW, January 2004.

147 **"those who aren't so good are actually discouraged":** From phone interview with John Birkbeck, conducted by CW, January 2004.

147 **twenty percent of elementary school kids have no PE at all:** From phone interview with Lorna Gillespie, conducted by CW, January 2004.

147–148 **one teacher covers every subject, from art to math to music to PE:** Ibid.

148 **average age is forty-seven and climbing:** Ibid.

148 **where kids get about thirty-five percent of their total calories:** From phone interview with Claude Marcus, conducted by CW, January 2004.

148 **obesity rates at these schools didn't budge, whereas rates in similar schools steadily climbed:** Ibid.

148 **announced a national strategy to encourage children and adults to exercise more and eat a more nutritious diet:** "King Launches Strategy to Tackle Obesity, Improve Nutrition and Increase Physical Activity," beehive.govt.nz (the official website of the New Zealand government), March 6, 2003.

149 **"maybe the food industry was partly responsible":** From phone interview with John Birkbeck, conducted by CW, January 2004.

149 **or any other positive change:** From phone interview with John Birkbeck, conducted by CW, January 2004.

149 **a bill to ban TV ads for high-fat or high-sugar foods during children's programs:** "Call for TV Food Ads Ban," BBC News, November 4, 2003.

149 **contains up to eleven ads for unhealthy foods:** Ibid.

149 **fell by the wayside in March 2004:** "Junk Food Ads Ban Is Shelved," *The Express*, March 4, 2004.

149 **after stirring up strong opposition from the food industry:** Jason Deans, "Jowell Attacked Over Junk Food Ads," MediaGuardian.co.UK, January 14, 2004.

149 **main cause of childhood obesity was a lack of exercise–not overeating:** "Junk Food Ads Ban Is Shelved," *The Express*, March 4, 2004.

149 **"what we have to judge is whether a ban would be appropriate":** Ibid.

150 **taking the lead in addressing nutritional problems in the developing world:** From phone interview with Monika Blossner, conducted by CW, January 2004.

150 **seek out a wider variety of foods that offer the right combination of nutrients:** Ibid.

150 **designed to prevent the worldwide spread of obesity, especially among children:** "Global Strategy on Diet, Physical Activity and Health," World Health Organization, May 22, 2004.

150 **wouldn't be legally binding in any country:** Fiona Fleck, "Top Health Officials Adopt Global Plan to Cut Obesity," *British Medical Journal*, May 29, 2004.

150 **encouraging the fast food industry to cut back on harmful ingredients such as trans fats:** "Global Strategy on Diet, Physical Activity and Health," World Health Organization, May 22, 2004. See also Peter Ford, "Foes of 'Globesity' Run Afoul of Sugar's Friends," *Christian Science Monitor*, February 19, 2004.

150 **a previous WHO report that called for people to get no more than ten percent of their calories from sugar:** Report 916, World Health Organization, April 2003. The Sugar Association attacked 916 in their own report entitled "Sound Science and Prospects for Sugar Consumption," *Agricultural Outlook Forum*, February 20, 2004.

151 **Health experts lauded the proposal:** Rob Stein, "U.S. Says It Will Contest WHO Plan to Fight Obesity But Claim of Faulty Science Is Rejected by Nutritionists," *Washington Post*, January 16, 2004.

151 **"very much in favor" of the Global Strategy:** Wendy Lubetkin, "Thompson Welcomes Global Strategy on Diet and Health," United States Embassy, Japan.

151 **a thirty-page letter addressed to the director general of the World Health Organization:** As of July 2004, this letter was available at http://www.commercialalert.org/bushadmincomment.pdf. The letter is discussed at length in "Secret Document Shows Bush Administration Effort to Stop Global Anti-Obesity Initiative," *Commercial Alert*, January 15, 2004.

151 **not supported by science, at least in the eyes of the United States:** From phone interview with Bill Pierce, conducted by CW, February 2004.

151 **American Psychological Association:** "Television Advertising Leads to Unhealthy Habits in Children; Says APA Task Force," American Psychological Association, February 23, 2004.

151 **Kaiser Family Foundation:** "The Role of Media in Childhood Obesity," the Henry J. Kaiser Family Foundation, February 2004.

151 **raised the risk of obesity by sixty percent:** David Ludwig, et al., "Relation Between Consumption of Sugar-Sweetened Drinks and Childhood Obesity: A Prospective, Observational Analysis," *The Lancet*, February 17, 2001.

151 **from government-imposed solutions to "personal responsibility":** Rob Stein, "U.S. Says It Will Contest WHO Plan to Fight Obesity But Claim of Faulty Science Is Rejected by Nutritionists," *Washington Post*, January 16, 2004.

151 **including inserting the words "personal" or "individual" nine times:** Michele Simon, "Washington Wages Super-Size Effort to Weaken WHO Report," Pacific News Service, May 7, 2004.

152 **"incorporated industry language into its critique practically word for word":** From email interview with Marion Nestle, conducted by CW, June 2004.

152 **"but they provided the motivation":** From phone interview with Neville Rigby, conducted by CW, January 2004.

152 **"they could do a lot better than throwing their hands up and saying it's personal responsibility":** From phone interview with Michael Lowe, conducted by CW, February 2004.

152 **"Nobody can show me any evidence of that":** From phone interview with Bill Pierce, conducted by CW, February 2004.

152 **with few concessions to the Americans:** Fiona Fleck, "WHO Resists Food Industry Pressure on Its Diet Plan," *British Medical Journal*, April 24, 2004.

152 **improve the quality of their foods and to limit marketing to children:** "Global Strategy on Diet, Physical Activity and Health," World Health Organization, May 22, 2004.

152 **made almost no mention of the WHO report that called for people to get fewer than ten percent of their calories from sugar:** Fiona Fleck, "WHO Resists Food Industry Pressure on Its Diet Plan," *British Medical Journal*, April 24, 2004.

152 **merely recommended "limiting sugar":** "Global Strategy on Diet, Physical Activity and Health" World Health Organization, May 22, 2004.

153 **"who are facing the issue of overweight and obesity":** J. Zarocostas, "WHA Adopts Landmark Strategy on Diet, Health," *The Lancet*, May 29, 2004.

153 **doesn't think that the WHO's strategy will have any effect on U.S. policy:** From email interview with Marion Nestle, conducted by CW, June 2004.

153 **"shameful actions in protecting the health of sugar producers over the health of children":** Ibid.

Chapter 7: Parents: What Helps, What Hurts

Page

155 **tour of Butler, Alabama, is a six-stoplight trip:** CW visited Butler in April 2004.

156 **dropped fifty pounds and six dress sizes while growing two inches taller, and she's still heading in the right direction:** From in-person interview with Amber Kearley and Susan Ryals, conducted by CW, April 2004.

156 **closest thing to a walking path in Butler is a dirt road surrounding an abandoned clothes factory:** Ibid.

156 **"The more you feed your children, the better mom you are":** From phone interview with Melinda Sothern, conducted by CW, May 2003.

157 **Georgia Pacific paper mill:** Ibid.

158 **"I would eat a bunch of chips and then move on to something else":** From in-person interview with Amber Kearley, conducted by CW, April 2004.

158–159 **"who were overweight and had health problems like diabetes or heart trouble":** From phone interview with Susan Ryals, conducted by CW, June 2003.

159 **worried about her son Jake's weight pretty much since the day of his birth:** From phone interview with Karen Rodgers, conducted by EH, July 2003.

159 **Jake has had the good fortune to escape cruel teasing by other kids:** Ibid.

159 **kids to take full responsibility for their weight:** L. H. Epstein, et al., "Ten-Year Follow-Up of Behavioral, Family-Based Treatment for Obese Children," *Journal of the American Medical Association*, November 21, 1990.

160 **landmark study in the *Journal of the American Medical Association* in 1990:** Ibid.

160 **the rates of obesity had climbed significantly in the other groups:** Ibid.

160 **and that includes doctors, nutritionists, friends at school, and even the child herself:** Moria Golan and Scott Crow, "Parents Are Key Players in the Prevention and Treatment of Weight-Related Problems," *Nutrition Reviews,* January 2004.

160 **astonishing power to shape their kids' eating habits:** From phone interview with Moria Golan, conducted by CW, April 2004.

160 **a study that underscored her message:** Moria Golan and Scott Crow, "Targeting Parents Exclusively in the Treatment of Childhood Obesity: Long-Term Results," *Obesity Research,* February 2004.

160 **the kids who stayed home had moved even closer to their ideal weight:** Ibid.

160 **girls in this group regularly forced themselves to vomit after their eating sprees:** Ibid.

160 **because they weren't denying themselves anything:** From phone interview with Moria Golan, conducted by CW, April 2004.

160 **enlist the support of parents and siblings from the very beginning:** From phone interview with Melinda Sothern, conducted by CW, May 2003.

160 **"your responsibility to make things better, but it's not your fault":** Ibid.

164 **single mother was convicted of misdemeanor child neglect after her 672-pound, thirteen-year-old daughter died of heart failure:** "Fat Child's Mother Guilty of Neglect," BBC News, January 10, 1998.

164 **"is going to help parents who are trying to help kids lose weight":** From phone interview with Melinda Sothern, conducted by CW, May 2003.

164 **parents know only too well what their children are going through:** From phone interview with Martha Lee Polatta, conducted by CW, May 2003.

165 **unless the child seems lethargic or has an insatiable appetite:** From phone interview with Melinda Sothern, conducted by CW, December 2001.

165 **(Insatiable appetite in a toddler:** Prader-Willi Association USA, Prader-Willi Syndrome Basic Facts, July 2004.

165–166 **kids and their parents line up to get weighed before a Shapedown meeting begins:** EH observed a Shapedown session in Palo Alto, California, July 2003.

166 **she encourages the parents to stay positive, to keep a sense of humor–and to be realistic:** From in-person interview with Anne Chasson, conducted by EH, July 2003.

167 **launch an all-out counterattack against a fattening culture:** From phone interview with Jennifer Fisher, conducted by CW, April 2004.

167 **strictly forbid certain types of food may actually be doing more harm than good:** Ibid.

167 **ate twenty-five percent less when allowed to chose their own portion sizes:** Alfredo Flores, "Larger Portions May Lead Children to Overeat," Agricultural Research Service, July 16, 2003.

167 **snack when they aren't hungry if parents take strict control over meals:** Jennifer Fisher and Leann Birch, "Eating in the Absence of Hunger and Overweight in Girls from 5 to 7 Years of Age," *American Journal of Clinical Nutrition*, 2002.

167 **"when you don't want kids to eat something, you don't give it to them":** From phone interview with Jennifer Fisher, conducted by CW, April 2004.

167–168 **studying preschoolers is a tough way to make a living:** Jennifer Fisher and Leann Birch, "Restricting Access to Palatable Foods Affects Children's Behavioral Response, Food Selection, and Intake," *American Journal of Clinical Nutrition,* 1999.

168 **diet pop for their chubby children and ice cream for themselves:** From phone interview with Melinda Sothern, conducted by CW, May 2003.

168 **"Pick the easiest things to change first":** From phone interview with Judith Levine, conducted by EH, August 2003.

169 **doesn't feel she can let her guard down for one moment:** From phone interview with Tracy Graham, conducted by CW, June 2004.

169 **parents in this situation can take a different approach:** From phone interview with Moria Golan, conducted by CW, April 2004.

169 **basic knowledge of nutrition, an array of healthy food options, and a few positive role models:** From phone interview with Jennifer Fisher, conducted by CW, April 2004.

170 **put on weight seemingly overnight:** From in-person interview with Molly and Libby Markus, conducted by CW, May 2004.

171 **"set up the environment in such a way that there's not a constant struggle":** Ibid.

171 **fried chicken with a side of macaroni and cheese, and a two-liter bottle of Coke:** From in-person interview with Susan Ryals, conducted by CW, April 2004.

172 **parents should encourage them to "eat healthy":** From phone interview with Moria Golan, conducted by CW, April 2004.

172 **Oprah Winfrey conducted a "Family Dinner Experiment":** Martha Marino and Sue Butkus, "Background: Research on Family Meals," a paper from Washington State University Nutrition Education.

172 **kids who regularly ate at home with their families had the healthiest diets:** M. W. Gillman, et al., "Family Dinner and Diet Quality Among Older Children and Adolescents," *Archives of Family Medicine,* March 2000.

172 **more fiber, iron, calcium, folic acid, and vitamins C, E, B6, and B12:** Ibid.

173 *Unless otherwise noted, information in "Healthy Eating at Every Age" comes from Lisa Tartamella.*

173 **easily absorbed iron, calcium:** Rebecca Williams and Isadora Stehlin, "Breast Milk or Formula: Making the Right Choice for Your Baby," *FDA Consumer Magazine*, June 1996.

173 **breastfeeding may help lower the risk of childhood obesity years down the road:** M. W. Gillman, et al., "Risk of Overweight Among Adolescents Who Were Breastfed as Infants," *Journal of the American Medical Association*, May 16, 2001.

173 **have more control over their own calorie intake:** Ibid.

173 **babies weigh at least thirteen pounds or have doubled their birthweight and can sit up with a little bit of help:** R. L. Duyff, *American Dietetic Association Complete Food and Nutrition Guide,* 2nd ed., 2002.

173 **easily tolerated and helps babies to meet their iron needs:** Mary Story, et al., *Bright Futures in Practice: Nutrition*, National Center for Education in Maternal and Child Health, 2000.

173 **single puréed vegetables and fruits and then strained meats and poultry:** Ibid.

173 **raw carrots, raisins, and whole grapes:** "What Do I Need to Know to Feed My Baby Safely?" University of Michigan Health System, October 2003.

173 **The American Academy of Pediatrics urges parents not to serve these foods until a child turns four:** Ibid.

174 **limiting daily intake to six ounces:** "Feeding Your Child," University of Michigan Health System, May 2003.

174 **not the time to give your child skim milk or nonfat cheese:** "Toddler Diet Lacking in Nutrition, Fat," Yale-New Haven Hospital Healthlink, August 24, 2000.

174 **crucial for the development of the brain and nervous system:** Ibid.

174 **leading source of saturated fat in kids' diets:** "The 1% or Less School Kit," Center for Science in the Public Interest.

174 **dark green, deep yellow, and orange:** Cecilia Wilkinson, et al., "Trends in Food and Nutrient Intakes by Children in the United States," *Family Economics and Nutrition Review*, 2002.

174 **Surveys by the USDA show that teenange girls:** "What and Where Our Children Eat–1994 Nationwide Survey Results," USDA and the Agricultural Research Service, April 18, 1996.

174–175 **eighty percent of the recommended daily allowance (RDA) of magnesium and calcium:** "Continuing Survey of Food Intakes by Individuals (CSFII) 1994–1996," National Research Council, August 1998.

175 **boys and girls tend not to get enough zinc and iron:** Ibid.

175 **Calcium:** R. L. Duyff, *The American Dietetic Association's Complete Food and Nutrition Guide*, 2nd ed., 2002.

175 **Magnesium:** Ibid.

175 **Zinc:** Ibid.

175 **Iron:** Ibid.

175 **on average it's higher in fat and calories than food prepared at home:** Joanne Guthrie, et al., "Role of Food Prepared Away from Home in the American Diet, 1977–78 Versus 1994–1996: Changes and Consequences," *Journal of Nutrition Education and Behavior,* May/June 2002.

175 **family meal was a two-hour project:** R. L. Duyff, *American Dietetic Association Complete Food and Nutrition Guide*, 2nd ed., 2002.

179 **reward or punish your child may lead to behavior problems related to food and mealtimes:** Bettylou Sherry, et al., "Attitudes, Practices, and Concerns About Child Feeding and Child Weight Status Among Socioeconomically Diverse White, Hispanic, and African-American Mothers," *Journal of the American Dietetic Association*, February 2004.

180 **love affair with these foods that will be nearly impossible to break:** Leann Birch, "Development of Food Preferences," *Annual Review of Nutrition*, 1999.

180 **only purėed fruits might lead him to believe that all foods should be sweet:** R. L. Duyff, *The American Dietetic Association's Complete Food and Nutrition Guide*, 2nd ed., 2002.

180 **yellow first, orange and pale green next, dark green and red last:** Ibid.

180 **stomach may not be equipped to handle their high nitrate content:** *Pediatric Nutrition Handbook,* 4th ed., American Academy of Pediatrics, 1998.

182 **"anything to get them away from the TV":** From phone interview with Melinda Sothern, conducted by CW, December 2001.

182–183 **six times as likely to be active if their parents are active:** L. L. Moore, et al., "Influence of Parents' Physical on Activity Levels of Young Children," *Journal of Pediatrics*, February 1991.

183 **kids will gladly exercise if they see a payoff:** From phone interview with James Sallis, conducted by CW, June 2004.

184 **overweight children have a much lower "exercise tolerance" than other kids:** From phone interview with Melinda Sothern, conducted by CW, December 2001.

184 **ride an exercise bicycle with the resistance turned down to zero for twenty minutes twice a week:** Ibid. See also M. S. Sothern, et al., *Trim Kids*, HarperResource, 2002.

185 **the "off" button on the remote can be one of your most powerful tools as a parent:** From phone interview with James Sallis, conducted by CW, June 2004.

185 **a TV in a child's bedroom is a strong risk factor:** B. A. Dennison, et al., "Television Viewing and Television in Bedroom Associated with Overweight Risk Among Low-Income Preschool Children," *Pediatrics*, June 2002.

185 **convinced a group of kids to cut their TV time by one third to one fourth:** T. N. Robinson, "Reducing Children's Television Viewing to Prevent Obesity: A Randomized Controlled Trial," *Journal of the American Medical Association*, October 27, 1999.

185 **cut Jake's television watching down to twenty hours a week:** From phone interview with Karen Rodgers, conducted by EH, July 2003.

186 **"I feel bad about that":** Ibid.

186 **Amber has plenty of chances to walk:** From in-person interview with Amber Kearley and Susan Ryals, conducted by CW, April 2004.

Chapter 8: A New Deal for Kids

Page

187 **in grocery stores, department stores, fast food restaurants, teachers' lounges, and, of course, airplanes:** "Chronology of Significant Developments Related to Smoking and Health," Centers for Disease Control and Prevention, May 2004. Arizona became the first state to restrict smoking in public places in 1973.

188 **"First, we have to work hard to spread the gospel":** From a speech by U.S. Secretary of Health and Human Services Tommy Thompson, given at Time/ABC News Summit on Obesity, Williamsburg, Virginia, June 2, 2004.

188 **two kids find a disembodied belly:** Department of Health and Human Services public education advertising campaign.

188 **State governments generally haven't distinguished themselves:** A. Cotton, "Report Card: State Efforts to Control Obesity," news release, University of Baltimore.

189 **lauded voluntary food industry changes:** From a speech by U.S. Secretary of Health and Human Services Tommy Thompson, given at Time/ABC News Obesity Conference, Williamsburg, Virginia, June 2, 2004.

189 **"We sounded the alarm":** Ibid.

189 **"everybody standing around the campfire":** Michele Simon, *Informed Eating Newsletter*, June 2004, quoting a talk by Kelly Brownell from the Time/ABC News Summit on Obesity, Williamsburg, Virginia, June 2, 2004.

189 **"I have a lot of respect for Secretary Thompson":** From David Satcher's keynote address, "Condition Critical: Covering Children's Health," given at the Casey Journalism Center on Children and Families, March 14, 2004.

189–190 **"The anthrax scare may have affected six individuals":** From an email interview with Roberto Treviño, conducted by EH, June 2004.

190 **"Quality, daily physical education":** From CDC report, "Promoting Better Health for Young People Through Physical Activity and Sports: A Report to the President," Fall 2000.

190 **the country needs after-school programs:** Ibid.

190 **strict Texas policy for competitive foods:** Texas Public School Nutrition Policy, Texas State Department of Agriculture.

191 **"If we're going to have nutritional standards":** From phone interview with Margo Wootan, conducted by EH, June 2004.

192 **"Give me a million mad moms":** Michele Simon, *Informed Eating Newsletter*, June 2004, quoting a talk by Susan Combs from the Time/ABC News Obesity Conference, Williamsburg, Virginia, June 2, 2004.

193 **"A lot of times it takes someone":** From phone interview with Margo Wootan, conducted by EH, June 2004.

193 **schools in Singapore and Hong Kong:** T. Lobstein, L. Baur, and R. Uauy for the IASO International Obesity Task Force, "Obesity in Children and Young People: A Crisis in Public Health," *Obesity Reviews*, May 2004.

193 **Treviño, who works closely:** From an email interview with Roberto Treviño, conducted by EH, June 23, 2004.

193 **"If schools are going to help them hook up":** From a phone interview with Francine R. Kaufman, conducted by LU, June 2004.

194 **many doctors feel frustrated:** E. Jelalian, et al., "Survey of Physician Attitudes and Practices Related to Childhood Obesity," *Clinical Pediatrics*, April 2003.

194 **tested the nutrition knowledge of doctors:** Peter Jaret, "What Medical School Didn't Teach You About Nutrition," *Hippocrates*, March 1999.

194 **'Can I send the patient to a nutritionist?':** From in-person interview with Roberto Treviño, conducted by EH, May 19, 2004.

194 **they felt torn and guilty:** Jill Andresky Fraser, *White-Collar Sweatshop*, 201.

195 **nearly seventy percent of them say they don't have enough time:** Jared Sandberg, Cubicle Culture column, *Wall Street Journal*, June 30, 2004.

196 **"Having a sense of comfort":** From an in-person interview with Joan Miller, conducted by EH, May 17, 2004.

196 **Only one supermarket serves West Oakland's:** From Peoplesgrocery.org.

197 **"At first glance, you could dismiss the community garden":** Linnea Due, "A Patch of Green," *East Bay Express*, February 20, 2002.

197 **Now Waters is taking the concept one step further:** "Memorandum of Understanding Between Berkeley Unified School District and Chez Panisse Foundation for a School Lunch Curriculum Initiative," passed by the school board, June 2004.

198 **"Kids need to find these foods to be delicious":** Peggy Orenstein, "Food Fighter," *New York Times*, March 7, 2004.

198 **up and running in 400 school districts:** Ibid.

198 **has more than 80,000 members worldwide:** From slowfood.com.

198 **leading lights gathered for a lecture:** Description by EH, who attended the event.

199 **regular family meals improved language skills:** Martha Marino and Sue Butkus, "Background: Research on Family Meals," a paper from Washington State University Nutrition Education.

Bibliography

America's Children: Key National Indicators of Well-Being 2003, Federal Interagency Forum on Child and Family Statistics, Washington, D.C., 2003.

Bowman, Shanthy A., Steven L. Gortmaker, Cara B. Ebbeling, Mark A. Pereira, and David S. Ludwig, "Effects of Fast-Food Consumption on Energy Intake and Diet Quality Among Children in a National Household Survey," *Pediatrics*, January 2004.

Bray, George A., Samara Joy Nielsen, and Barry M. Popkin, "Consumption of High-Fructose Corn Syrup in Beverages May Play a Role in the Epidemic of Obesity," *American Journal of Clinical Nutrition*, April 2004.

Brownell, Kelly D., and Katherine Battle Horgen, *Food Fight*, New York, McGraw-Hill, 2004.

Carmona, Richard, *The Growing Epidemic of Childhood Obesity in the United States,* Department of Health and Human Services, Washington, D.C., 2004.

Chakravarthy, M. V., and F. W. Booth, "Inactivity and Inaction: We Can't Afford Either," *Archives of Pediatric and Adolescent Medicine*, August 2003.

Consequences of Overweight in Children and Adolescents, Centers for Disease Control and Prevention, Atlanta, Georgia, 2002.

Dale, D., and C. Corbin, "Restricting Opportunities to Be Active During School Time: Do Children Compensate by Increasing Physical Activity Levels After School?" *Research Quarterly for Exercise and Sport*, 2000.

De Onis, Mercedes, and Monika Blossner, "Prevalence and Trends of Overweight Among Preschool Children in Developing Countries," *American Journal of Clinical Nutrition*, October 2000.

Dennison, B. A., T. A. Erb, and P. L. Jenkins, "Television Viewing and Television in Bedroom Associated with Overweight Risk Among Low-Income Preschool Children," *Pediatrics*, June 2002.

Dietz, William, "Overweight in Childhood and Adolescence," *New England Journal of Medicine*, February 26, 2004.

Drewnowski, Adam, "Poverty and Obesity: The Role of Energy Density and Energy Costs," *American Journal of Clinical Nutrition*, January 2004.

Ebbeling, Cara, D. B. Pawlak, and D. S. Ludwig, "Childhood Obesity: Public-Health Crisis, Common Sense Cure," *The Lancet*, August 2002.

Ewing, R., T. Schmid, R. Killingsworth, A. Zlot, S. Raudenbush, "Relationship Between Urban Sprawl and Physical Activity, Obesity and Morbidity," *American Journal of Health Promotion,* September/October 2003.

Fighting Hunger Today Could Fight Obesity Tomorrow, UN Food and Agriculture Organization, February 11, 2004.

Food Choices at School: Risks to Child Nutrition and Health, Call for Action, Democratic Staff of the Senate Committee on Agriculture, Nutrition, and Forestry, Washington, D.C., May 18, 2004.

Forshee, R. A., and M. L. Storey, "Total Beverage Consumption and Beverage Choices Among Children and Adolescents," *International Journal of Food Science and Nutrition*, July 2003.
Fraser, Jill Andresky, *White-Collar Sweatshop*, New York, W. W. Norton, 2001.
Global Strategy on Diet, Physical Activity and Health: Report by the Secretariat, World Health Organization, April 17, 2004.
Gortmaker, S. L., A. Must, A. M. Sobol, K. Peterson, G. A. Colditz, and W. H. Dietz, "Television Viewing as a Cause of Increasing Obesity Among Children in the United States, 1986–1990," *Archives of Pediatrics and Adolescent Medicine*, 1996.
Healthy People 2010, 2nd ed., U.S. Department of Health and Human Services, Washington, D.C., 2000.
Healthy Schools for Healthy Kids, Robert Wood Johnson Foundation, Princeton, New Jersey, 2003.
Insulin Resistance and Pre-Diabetes, National Diabetes Information Clearinghouse, Bethesda, Maryland, 2004.
Jaret, Peter, "What Medical School Didn't Teach You About Nutrition," *Hippocrates*, March 1999.
Levine, Judith, and Linda Bine, *Helping Your Child Lose Weight the Healthy Way,* New York, Kensington, 1996, 2001.
Lobstein, T., L. Baur, and R. Uauy, "Obesity in Children and Young People: A Crisis in Public Health," *Obesity Reviews*, 2004.
Lowe, Michael, "Self-Regulation of Energy Intake in the Prevention and Treatment of Obesity: Is It Feasible?" *Obesity Research*, October 2003.
Ludwig, David S., Karen E. Peterson, and Steven L. Gortmaker, "Relationship Between Consumption of Sugar-Sweetened Drinks and Childhood Obesity: A Prospective, Observational Analysis," *The Lancet*, February 17, 2001.
Molnar, Alex, *No Student Left Unsold*, Commercialism in Education Unit, Arizona State University, October 2003.
Nestle, Marion, *Food Politics,* Berkeley, University of California Press, 2002.
Pediatric Nutrition Handbook, 4th ed., Elk Grove Village, IL, American Academy of Pediatrics, 1998.
Popkin, Barry, and Samara Nielsen, "The Sweetening of the World's Diet," *Obesity Research*, November 2003.
The Role of Media in Childhood Obesity, Henry J. Kaiser Family Foundation, Menlo Park, California, February 2004.
Soft Drinks in Schools, Elk Grove Village, IL, American Academy of Pediatrics, January 2004.
Statistics Related to Overweight and Obesity, National Institute of Diabetes & Digestive & Kidney Disease, Weight Control Information Network, Bethesda, Maryland, 2003.
Texas School Nutrition Policy, Texas Department of Agriculture, Austin, Texas, 2004.
2000 California High School Fast Food Survey: Findings and Recommendations, Public Health Institute, Sacramento, California, 2000.

Suggested Reading

Betschart-Roemer, Jean, *Type 2 Diabetes in Teens*, New York, John Wiley, 2002.
Cunningham, Marion, *Lost Recipes: Meals to Share with Friends and Family,* New York, Alfred A. Knopf, 2003.
Nissenberg, Sandra K., Margaret L. Bogle, and Audrey C. Wright, *Quick Meals for Healthy Kids and Busy Parents*, New York, John Wiley, 1995.
Satter, Ellen, *How to Get Your Kids to Eat . . . But Not Too Much*, Boulder, Bull Publishing, 1987.
Sears, William, Peter Sears, and Sean Foy, *Dr. Sears' Lean Kids*, New York, New American Library, 2003.
Schlosser, Eric, *Fast Food Nation,* New York, Houghton Mifflin, 2001.
Sothern, Melinda S., T. Kristian von Almen, and Heidi Schumacher, *Trim Kids*, New York, HarperCollins, 2001.

Resources

American Dietetic Association: the country's largest organization of nutrition professionals; website includes tips for consumers as well as how to find a nutritionist in your area. www.eatright.org.

American Obesity Association (formerly American School Food Service Association): focuses on improving public policy and changing perceptions concerning obesity. www.obesity.org.

Boys and Girls Clubs of America: offers sports, fitness, and recreation programs that foster physical activity and social skills, including teen sports clubs and inner-city baseball leagues. www.bgca.org/programs/sportfitness.asp.

California Project LEAN (Leaders Encouraging Activity and Nutrition): under the California Department of Health Services; operates such programs as Food on the Run, dedicated to increasing healthy eating and physical activity among adolescents, as well as working toward healthier foods in schools. www.californiaprojectlean.org.

CANFit (California Adolescent Nutrition and Fitness Program): based in Berkeley, California, creators of P.H.A.T (Promoting Healthy Activities Together). www.canfit.org.

CATCH (Coordinated Approach to Childhood Health): Texas-based school program encouraging good nutrition and fitness; website has recommendations for families as well as educators and child nutrition services. http://www.sph.uth.tmc.edu/catch.

The Center for Health and Health Care in Schools: a policy and program resource center at the George Washington University School of Public Health and Health Services; provides numerous links to a variety of organizations dealing with childhood obesity. www.healthinschools.org/sh/obesity.asp.

Center for Informed Food Choices: advocates for eating a whole foods, plant-based diet and educates about the politics of food; publishes *Informed Eating*, a newsletter of food politics and analysis. www.informedeating.org.

Center for Science in the Public Interest: based in Washington, D.C., a nutrition advocacy organization. CSPI publishes the *Nutrition Action Healthletter*, a wealth of information on healthy eating as well as foods to avoid; rates the fat and sugar content in foods served by many of the nation's chain restaurants. www.cspinet.org.

Commercial Alert: the email newsletter provides information related to commercialism and marketing and advertising successes, especially marketing aimed at children. www.commercialalert.org.

The Edible Schoolyard at Martin Luther King Middle School, Berkeley, California: dedicated to integrating an organic garden tended by students into the school curriculum and lunch program. www.edibleschoolyard.org.

Federal 5 a Day program: under the Centers for Disease Control and Prevention, the program provides materials encouraging people to eat five servings of fruits and vegetables every day. http://www.cdc.gov/nccdphp/dnpa/5aday/.

National Association for Sport and Physical Education: develops and promotes physical activity programs for physical education teachers and coaches. http://www.aahperd.org/naspe/.

National Farm to School Program: connects school food services with local farms. www.farmtoschool.org.

National Institute of Diabetes & Digestive & Kidney Diseases, Weight Loss and Control Center: offers consumer advice and tips for weight management. http://www.niddk.nih.gov/health/nutrit/nutrit.htm.

Nutrition & Physical Activity: under the Centers for Disease Control and Prevention, provides reports and information for consumers on how to integrate better nutrition and physical activity. http://www.cdc.gov/nccdphp/dnpa/index.htm.

School Nutrition Association: a member organization for school food managers, formerly the American School Food Service Association, SNA's mission is to improve the availability and quality of school nutrition programs; website offers breaking news on school food policy as well as nutrition advice. www.asfsa.org.

Slow Food: dedicated to ecologically sound food production, home-cooked meals, and to "living a slower, more harmonious rhythm of life." www.slowfoodusa.org or www.slowfood.com.

SPARK P.E. program: a California-based program, evolved from a National Institutes of Health grant to create an elementary physical education program, now used by numerous elementary schools. www.sparkpe.org.

Strategic Alliance for Healthy Food and Activity Environments: a coalition of nutrition and physical activity advocates in California, based in Oakland. www.preventioninstitute.org/sa.

Texas State Department of Agriculture: the website includes the full text of the Texas School Nutrition Policy, effective August 2004, which can be found at: http://www.agr.state.tx.us/foodnutrition/policy/food_nutrition_policy.pdf.

Walk and Bike to School: to enhance kids' health, help the environment, and create safer routes to school. www.walktoschool-usa.org.

Walking School Bus: maintained by the Pedestrian Bicycle Information Center for the Partnership for a Walkable America; provides information and links for adults walking kids to and from school. www.walkingschoolbus.org.

YMCA Activate America: in partnership with government and research organizations, the initiative aims to improve the nation's health and fitness. www.ymca.net or http://www.ymca.net/activateamerica/activateamerica.jsp.

Acknowledgments

In the course of writing this book, we met many children and parents who amazed us with their determination and courage. They willingly shared their stories–hardships and all–and this book couldn't exist without them. We offer our deep thanks to Susan Ryals and Amber Kearley, Tracy and Josh Graham, Libby and Molly Markus, Sherri Jarrett and Blake Crenshaw, Karen and Jake Rodgers, Cheryl and Eva Afghani, Patricia Orozco and Yadira Renteria, Alma and Ameri Lopez, Carmen Miranda, and Clay Jones.

The book has been a team effort from the outset. Consumer Health Interactive Executive Editor Diana Hembree envisioned this book before we did. Her wisdom, inspiration, unflagging enthusiasm, and graceful editing saw us through this project from beginning to end. We are grateful to CHI Articles Editor Psyche Pascual for her help with reporting and editing and to CHI contributing writers Laurie Udesky and Paige Bierma for their assistance with research, reporting, and fact-checking. We also wish to thank Nancy Montgomery, Jennifer Robb, Benj Vardigan, and the rest of the CHI editorial staff and our other colleagues.

Our agent, Barbara Moulton, believed in the power of this story from the moment we pitched it, as did JoAnn Miller, our editor at Basic Books. Thanks also to Dan Segedin, Connie Matthiessen, Barbara Jamison, Diana Reiss-Koncar, Brady Kahn, and Elizabeth Bell.

Of the dozens of experts who gave us their time and insights, several deserve special mention. Joan Miller of the Bexar County Community Health Collaborative in San Antonio was an enthusiastic fount of information and sources. Melinda Sothern gave us several interviews that provided a foundation for the book. Jurgen Kraehmer, Roy Leon, and Jim Hinkle let us observe their gym classes and soak in their knowledge of the "New PE."

Gina Castro and Sally Cody piloted us to school cafeterias and physical education classes throughout San Antonio. Dr. Roberto Treviño and his staff generously connected us with children in his program. Dana Woldow, Caroline Grannan, and Linal Ishibashi helped us understand the mysteries of the school lunch program and showed us that just a few people can make a difference. We are also indebted to the Casey Journalism Center on Children and Families for its journalism fellowship program, which provided invaluable expertise and resources.

Finally, we'd like to thank our families. Blythe Woolston did especially fine work as an in-house (literally) editor and indexer. Robin Evans helped us polish many a phrase, even as she took over the crafting of those highly recommended family meals. Greg Kimmel offered enthusiasm and support throughout. All in all, our spouses and kids kept us going with their patience and encouragement. We are also truly grateful to our parents–Jan and Dee Woolston, Joanne Bliven and Michael Tartamella, and Mildred Herscher and the late Edward Herscher–for their lifelong support of our work.

Index

G

Q

R

S

W

Y

Z